Fostering Innovation and Entrepreneurship

Christian Schwarzkopf

Fostering Innovation and Entrepreneurship

Entrepreneurial Ecosystem and
Entrepreneurial Fundamentals in
the USA and Germany

Christian Schwarzkopf
Karlsruhe, Germany

Karlsruhe Institute of Technology, Germany, 2015

OnlinePlus material to this book can be available on
http://www.springer-gabler.de/978-3-658-13511-9

ISBN 978-3-658-13511-9   ISBN 978-3-658-13512-6 (eBook)
DOI 10.1007/978-3-658-13512-6

Library of Congress Control Number: 2016936553

Springer Gabler

Printed on acid-free paper

This Springer Gabler imprint is published by Springer Nature
The registered company is Springer Fachmedien Wiesbaden GmbH

Acknowledgements

I would like to thank my thesis advisor Professor Rothengatter for having given me the opportunity to write about this exciting topic, for his continuous and great support, as well as for his valuable feedback throughout the years of this dissertation.

I would also like to thank my Professors, Orestis Terzidis and Andreas Oberweis, for their valuable feedback and motivation.

Many thanks go to my family and their support, especially my father Professor Schwarzkopf for his intense feedback and valuable discussions.

Finally, I would like to thank all the persons who have helped me to master this thesis, by giving advice and feedback, participating in the surveys, and by providing me with insights – from an investor's perspective, from academia as well as from an entrepreneur's perspective.

Last but not least, thanks to my business partner Tim, who supported my aspiration to pursue this doctoral thesis throughout this long period.

Christian Schwarzkopf

Abstract

Entrepreneurship and innovation have been identified in politics and in industry as crucial elements for economic success. In this context, the importance of a successful entrepreneurial environment is mentioned, and sometimes the term ecosystem is also being used. Yet the fairly new term Entrepreneurial Ecosystem has not been sufficiently defined, analyzed and developed, in order to understand what it actually stands for. The major goal of this thesis is to develop a holistic concept of an Entrepreneurial Ecosystem – taking economic, social and personal aspects into account – and identify areas, in which Germany nowadays can improve and learn from the USA. In the beginning, the areas of innovation and entrepreneurship have been analyzed, showing that innovation is all about implementation of new things in a market, and entrepreneurship is about persons that are part of the innovation process, by starting business with an invention or a new business aspect that they bring into the economy, according to their own believes and goals.

Not everyone can become a successful entrepreneur, because the necessary inherent abilities cannot be learned or taught. However, these qualities alone are not enough to guarantee success. Success factors of startups and its entrepreneurs have been part of an analysis and have been supplemented by two surveys the author conducted among investors and entrepreneurs. The surveys especially revealed that self-motivation, and the combination of individual ability and skills of the entrepreneurs were considered by both groups as the most important success criteria.

A comparison between the United States and Germany showed that today the US indeed is a more entrepreneurial country and has had, on average over the last years, about 40 times more available venture capital than Germany. Its open culture, its recent 200+ years of immigration history, as well as its investor friendly tax policy and its entrepreneur friendly bankruptcy laws all together favor the foundation and growth of new ventures in the US.

As a role model for global entrepreneurship, the US success factors, as well as the literature research and the surveys have helped to identify the four important areas of an Entrepreneurial Ecosystem. These areas occur in a circular form creating 4 independent circles, starting from the personal characteristics in the center to the impersonal business qualities and activities on the outside parameter. Every circle has its own distinct qualities and characteristics. Among them are abilities and skills in the "Personal Circle", family and friends in the "Private Circle", higher education and training in the "Educational Circle" and business world and culture in the "Public and Business Circle". For the purpose of this study of entrepreneurial phenomena and their success relation, the focus is on Germany and its needs for improvement, which are necessary to overcome the under-financing of startups and to induce a more pro-entrepreneurial climate. Encouraging the legal person – company or corporation – to invest parts of their yearly profits into venture capital is one of the potential solutions. Furthermore, looking into how the German state can encourage and motivate more successful entrepreneurs, inventors as well as investors to participate in startup businesses and companies. Eventually, a more unified and business open-minded Europe creates one huge common domestic market, in which startups could grow as large and as fast as startups can in the USA.

Table of Contents

List of Tables

List of Figures

List of Figures

Abbreviations

Bn	Billion
BA	Business Angel
BRIC	Brazil, Russia, India and China
BVK	(Bundesverband Deutscher Kapitalbeteiligungsgesellschaften)
CRM	Customer Relationship Management
ERP	Enterprise Resource Planning
EUR	Euro (Currency)
EXIT	Sell of the complete or parts of the company, especially to return the money to the investors
GDP	Gross Domestic Product
IAS	International Accounting Standard
IP	Intellectual Property
IPO	Initial Public Offering
K	Thousand
M	Million
OCEAN	Openness, Conscientiousness, Extraversion, Agreeableness, Neuroticism
PE	Private Equity
PMI	Project Management Institute
ROI	Return on Invest(ment)
SME	Small Medium Enterprise
SWOT	Strengths Weaknesses Opportunities Threats
T	Trillion
TEA	Total Early-Stage Entrepreneurship Activity, age 18-64 in the first 3.5 years
US/USA	United States of America
USD	US Dollar (Currency)
VC	Venture Capital

1 Introduction

1.1 A Rationale for Innovation & Entrepreneurship

Innovation can change our lives extensively, regardless of whether it is the food business, the medical sector, transportation, the energy industry or even information itself. Actual improvements and new introduced products have led to a longer life expectation, higher standard of living and a variety of options already available for many people on our planet. We can travel the world; indulge ourselves in leisure activities and use high-tech equipment like a smart phone to call a friend from the top of a mountain, while measuring the speed and height distances. Many things we take very often for granted, without recognizing the tremendous development behind it made by mankind. We might not need all of these innovations in our lives nor do they come without problems, but we have to embrace the chances and challenges to move on. We need new medical treatment to heal deadly diseases; new technology is necessary for clean energy production, enhanced water treatment and food production in order to feed billions of people in the future.

Since our society has become so exceptionally depended upon technology, there is no alternative to the continuous quest for more and better solutions in our work and private life. With the incredible high debts of almost every large nation, we are actually doomed to grow and expand globalization efforts. Yet this growth is closely connected with innovation, because our planet offers limited resources that need to be more intelligently used, reused as well as rebuild. This can only be done through further innovation, which means products, services or methods that find an implementation in practice. And this necessary growth is never over when seeing it on an international scale. China and India for example, with a population of more than one billion inhabitants each, are still far away from the standards of living of their European or US counterparts, thus need nearly endless growth to close the gap. They are a symbol and hope for an ongoing growth. Furthermore, founding new companies is also the best way to create jobs all over the world and thus offer personal freedom, peace and access to a higher standard of living.

Since 1980, America's Fortune 500 companies for example have cut more than 5 million jobs, whereas new small and medium companies, thereafter, have created more than 34 million jobs (Kuratko 2005). These founders, or entrepreneurs as one can also call them, stimulated the growth of the economy and continue to shape the globalized world. This globalization and economic growth will help nations to prosper and hopefully live in peace (Jones 2013; Marquardt 2005). It is also the author's opinion that people, who have jobs and who can create their future, are less inclined to go to war or become dangerous fanatics. The more they have sometimes to lose, the more they care about balancing the relations with others in order to maintain their status. Certainly, an imbalance by too much wealth, power and greed can also lead to acts of war and injustice. It is more than 2000 years ago when Aristotle already proclaimed in *The Nichomean Ethics* the golden mean and not to pursue anything to an excess – not even good fortune (Aristotles 350AD, p.VII.12.).

Despite the social and political change through economic developments, which is stimulated by innovation, new technology has also aided nations in their quest for political and personal freedom. Smart phones on the one hand, combined with social network services like Twitter on the other hand are one of the most recent examples of informing the world about discrimination and terror, thus leading to the "Arab Spring" in Tunisia, Egypt and other Middle East Nations.

New technology does not have to be life threatening. Which person really wants to live without the communication or other new technology we use every day? Listening to music, watching TV, taking a car, flying in a plane, talking on and using a smartphone, having a refrigerator, surfing the Internet and many other technology based innovations.

All this shows that innovation is one of the key leverages for change; a question, however, still remains: Who drives the innovation and brings ideas as well as inventions into the markets? Originally, individuals, who have invented products, do not necessarily need to introduce them to the public. This requires another person to bridge the gap between supply and demand. It involves someone, who sees opportunities and seeks to achieve them with new methods – a classic situation for businesspersons not inventors. This businessperson, who does not wait, if inventions or ideas find their way to the customer, but finds the way by him or herself. He or she is someone, who might need to break rules and stand-alone against the opinion of others. Innovation is in many cases hard to achieve and, thus, needs to be rewarded with extra benefits, due to the risk of failure entering a market with a new product or service. Usually this is connected with financial rewards. However, more social entrepreneurs can be found (The Economist 2006), who are mainly compensated by the positive benefits they bring to society and not themselves.

Eventually, we need to embrace both, the inventors and entrepreneurs, in order to bring more innovation to society and think of ways to support and encourage their aspirations.

The thesis will especially cover various US and German market aspects such as venture capital, history, geography, culture & society, economic markets, government regulations, character traits and top performing companies – with its research results being of use and application in other countries.

1.2   Innovation and Entrepreneurial Background

Invention, entrepreneurship and innovation in general come from very different social, academic or business backgrounds. Some entrepreneurs have never finished high school like Germany's Rheinhold Würth, who took over the father's little supply store after his death and expanded it extensively to a global billion-dollar company. Others like the Americans Bill Gates of Microsoft or Mark Zuckerberg of Facebook never finished university, but went on building up their company. The British Richard Branson never even started at a university, and still was very innovative and successful in founding and developing the Virgin Group. Larry Page, founder of Google, obtained a Bachelor degree in Engineering and a Masters Degree in Computer Science and before finishing his PhD at Stanford University founded Google.

The goal of this thesis is, therefore, to identify the influencing factors that lead to success. Due the author's proximity and passed closed connections to the university, some close relations of innovation coming from an academic influence are inevitable. That could range from a technology being developed at a university to close contacts with young entrepreneurs that have studied at university later founded companies.

Since universities and research foundations, however, have a long history of bringing out new companies and are deeply involved in technology development, many ideas and inventions have their roots in academia. However, where implementation is lacking, innovation fails. One of the first hypotheses of this thesis is, therefore, that higher education is an essential element to the Entrepreneurial Ecosystem and must be constantly adjusted and improved (compare section on hypotheses 2.2).

Eventually, this thesis will expand to include a variety of influencing factors on entrepreneurship, aside from focusing on academia.

1.3   German & US Market

Germany has become one of the most successful and wealthy nations in this world. It dominates many industries like automotive, machinery or chemical and has invented and implemented many products. Nevertheless, especially during the last years, more and more questions are being raised in Germany, why countries like the US or South Korea lead in many modern industry sectors, like Digital Economy, Biotechnology or Entertainment products and devices, although these sectors do not require natural resources but smart, well educated and creative people. Even though Entrepreneurship, venture capital and startup incubators require no natural resources, the US still outperforms Germany in these areas – they are the benchmark for successful entrepreneurship in the world. Due to the fact that both countries are economic super powers and have produced hundreds of great entrepreneurs and thousands of companies in the past century, it has especially become important today, 2009-2014, to compare them in the field of entrepreneurship and venture capital. This up-to-date comparison between these two countries, hopefully, may give some decisive insights to which elements of an Entrepreneurial Ecosystem are important and what type of person it takes to be successful in the world of tomorrow.

1.4   An Entrepreneurial Ecosystem

The study of innovation and entrepreneurship as the major field of study requires the consideration of several important aspects, such as the process of innovation, its implementation, the financing and the existing market constellations and the influences of Academia as well as family, friends or co-workers. The major task of this thesis will be identifying, understanding and clustering these important surrounding factors that influence the founder and explain his or her success. These and other factors form the so-called Entrepreneurial Ecosystem. The thesis will further try to design this Entrepreneurial Ecosystem from a macro perspective, and thus rather developing a generally accepted nationwide ecosystem, in which other micro Entrepreneurial Ecosystem from the perspective of a certain founder or a certain region can fit in without any contradiction.

It is necessary to understand and describe, what is really meant by an Entrepreneurial Ecosystem on each level of perspective, and how it is connected and functioning, in order to derive and develop means of improvement. This approach goes along with maxims of many other business people like the former major of New York and successful entrepreneur Michael Bloomberg, who is known for his statement "*If you can't measure it, you can't manage it*" (M. Bloomberg 2013). Eventually, it shall be possible to understand the system's elements and dynamics in the entrepreneurial context and enrich it with new ideas and designs.

1.5   Motivation

It is obvious in the comparison to the United States, that Germany needs more entrepreneurs and people, who want to start their own business. The more people create their own business, the higher the chances are that many of these companies create value and bring innovation to our society. Therefore, Germany has to become an entrepreneurship friendly environment. It means that Germany has to be open for any kind of new company foundation, rather than only focusing on high-tech industry, which is a part of the German culture, where things have to be 100% correct and on the surface be useful as well meaningful. Nevertheless, some entrepreneurs have to try several startups, till they find the right one and become successful enough to implement something big, sustainable or important. Especially in the early stage of an idea or company, it is hard or impossible to judge, whether it will be a success and in which direction it will turn. That is why the Germans as a whole, the government and the business and research institutions need to embrace also small and simple ideas instead of the perfectly founded and highly developed concepts. They all need to be encouraged, to think "outside the box" and globally from the beginning, without automatically judging these progressive thinkers and daydreamers. For many industries the time is over, in which a company can slowly grow in its domestic market and expand gradually abroad. The world is too connected, and business people scout the world for new concepts and products to be copied or reshaped in another country. In this context McDougal (McDougal 1989) already spoke in the late 1980s about the development of international startups that saw their business and markets to be global from the beginning of their operations.

Thus, all modern entrepreneurs must adopt the characteristics of being open minded, thinking globally, while still performing and understanding their job profoundly.

The author believes that some of these kinds of entrepreneurs exist in Germany but need to be stimulated, trained and formed to meet the needs of the future. Entrepreneurs should also be publically and privately supported, thus creating more value and innovation for society. One key element necessary for this support lies in the improvement of the whole Entrepreneurial Ecosystem in Germany. The stakeholders, whether public or private, the customers and the business community as a whole need to be encouraged to understand the importance of the overall concept of entrepreneurship, and all must play a positive active part in its development.

2 Methodology

There are four main questions:

- Why do most recent successful Internet companies and innovation come from the USA?
- Why do US venture funds take the risk of investing much earlier in many unproven startup concepts at higher evaluations?
- Why is the US environment for startups so much more pro-entrepreneurial?
- Why are many German advisors engaged in the entrepreneurship field, even though they have no experience in it?

Finding answers to those questions, understanding the differences of the German and US environment for entrepreneurs and eventually, deriving potential improvements became the initial goal of this thesis. The idea was then to develop a concept that bears all the relevant aspects of entrepreneurship & innovation, which lead to success.

The first hypothesis of this paper is that these relevant aspects and the environment of the entrepreneur can help to explain, what is meant by the so-called Entrepreneurial Ecosystem.

2.1 Approach

The general approach of the thesis can be divided into seven major parts:

1) Getting a deeper understanding of the subject through a first literature research
2) Generating further hypotheses
3) Examining the present state-of-the-art research
4) Generating specific primary data with input from experts (surveys among investors and entrepreneurs)
5) Comparing the US and German entrepreneurship culture in detail
6) Defining a conceptual framework for the Entrepreneurial Ecosystem
7) Deriving improvement areas

At the start of generating ideas for this thesis in 2008, there was considerable information on startups' successes, but hardly any information on Entrepreneurial Ecosystems. 2-3 years later, more and more reports and papers came out trying to use and define the term Entrepreneurial Ecosystem, but have not reached a level of a comprehensive definition.

In the beginning, the research goal is to review the general opinions in Germany and the US ecosystem. And then, based upon the actual entrepreneurial experience of the author, generate his hypotheses.

The main primary and secondary data collection for the analysis shall consist of

- Secondary data through studies and comparisons of journals, books, magazines, newspapers, relevant website etc.
- Primary data through surveys among investors and entrepreneurs from various countries
- Primary data through interviews with experts – especially from the United States and Germany – such as investors, managers and academics

Eventually, the results of this analysis will help deriving and designing a holistic Entrepreneurial Ecosystem. The final step will then be to identify where improvements can be made and to propose potential solutions for them.

2.2 Hypotheses

After first research steps on the subject, it was possible to come up with the following hypotheses. There are general hypotheses with more abstract statements and very specific ones with concrete assertions for specific goals and results.

The initial hypotheses are the following:

- A generally accepted Entrepreneurial Ecosystem can be defined from a macro perspective.
- Improving parts of the Entrepreneurial Ecosystem can increase a founder's success chances.
- Success means fulfilling more than just financial aspects.
- Higher education and universities are part of the Entrepreneurial Ecosystem.
- The German national Entrepreneurial Ecosystem lacks behind in comparison to the US.
- An entrepreneurial venture is mistakenly viewed as a linear sequence of an chronological event; it should be seen as a circular development.

Basic assumptions

- Entrepreneurship is linked to innovation.
- Individuals and institutions can help startups to become more successful.
- The government can foster entrepreneurship.
- A key for today's business improvement and success lies in smart cooperation and internationalization.

It is however not the goal of this thesis to prove the hypotheses by developing a universally valid model. The goal is to find a comprehensive definition and concept for the Entrepreneurial Ecosystem, and thus being able to better understand entrepreneurship, innovation and their influencing factors. The hypotheses are helping guideposts in the process of this general goal.

3 Knowledge Base

3.1 History of Innovation

For many people the term innovation stands for progress, success and something new in our lives. Some might think of their first iPhone, others of solar panels on their roof. But innovation is so much more; it moves our society forward, but brings also new challenges along. It keeps mankind in a continuous competition for customer demand and creates as well as destroys jobs (Kollmann 2006).

Originally, the word derives from the Latin verb *innovare* meaning to create or renew something (Stowasser et al. 1998). The Merriam-Webster Dictionary defines it as the introduction of something new or a new idea, method or device (Merriam-Webster 2010b). The earliest sources of the word *innovatio* can be found in some Latin Church texts of Tertullian around 200 BC and Augustin around 400 BC with the meaning *renewal* and *change* (Müller 1997, p.9). When Dante was using the term *innovare* (Dante 13th century, p.303) in the early 14th century, he was talking about a renewal of his rimes. Machiavelli (Machiavelli 1513) however was referring with his *innovatori* for the first time about men, who stood for change and new things, which they wanted to implement. These thoughts have to be reexamined again in the chapter on Entrepreneurship 3.4. Shakespeare used the word *innovate* in the context of political change, somehow like Machiavelli did (Müller 1997, p.54). Today's understanding of innovation in a more technical and economic sense was mainly formed by Schumpeter (Schumpeter 1939) in the 20th century, and probably also influenced by Machiavelli. Often, inventions and innovations get mixed up or are used similar. However, modern research, especially Schumpeter, clearly distinguishes between an invention and innovation, with the former being part of the latter. The missing element is implementation in a market. Schumpeter sees innovation combining factors in a new way and bringing them to life or converting the invention into the market (Schumpeter 1939, pp.85–88).

Leonardo da Vinci is often regarded as the greatest inventor of all times, and he can serve to point out the difference between invention and innovation. Many of his inventions like airplane prototypes, tanks or robotic knights had never been implemented during his life, and thus never be in use (Davinci Innovations 2008). Centuries later, people may have been inspired by his concepts and inventions to implement them and make them an innovation. This is the essence of Schumpeter's thought, that it takes other people – entrepreneurs – to actually implement the new ideas or inventions (compare the following section on Defining Entrepreneurship & Entrepreneur 3.4).

A question is, how and when did this constant and systematic process of renewing and introducing new things, what we today call innovation, occurred for the first time? This form of modernization must have happened from the early days of mankind on. Whether it was the wheel or the invention of iron, it always moved society forward. But there might have been an invention and its introduction, which has triggered a chain event and professionalized the innovation process. One idea is that it could have started more than two thousand years ago, with the invention and introduction of a military product: the stirrup. The stirrup is said to be the first military innovation that led to a professionalized continuous development of armament and new products. Its first appearance was about 200 BC in India, however modern stirrup types can be first traced back to China in the 4th century AD (Dien 1986). It took further four hundred years, till it reached Europe and made the knights cavalry much stronger and preparing the ground for the dominance of the feudal class (White 1966). But the effect was even bigger, because this technical combat advantage inspired an on-going improvement, first in a military context, later it swept over to civilian areas. In addition, the stirrup stands for those inventions that lead to a leapfrog and not simple an efficiency improvement. Being implemented, such inventions lead to innovation that stands for a revolution and not evolution in their market or environs – similar to the invention of the wheel that has revolutionized civilian and military life. In contrast, a comparison of the stirrup with the chariot, which also improved warfare, shows that the chariot rather was an efficiency improvement. The high price, complexity and disadvantage in certain terrains of the chariot stands against the simplicity, low price and total appliance of the stirrup on all horses (Derby 2001), which led to its strong and complete market penetration, and therefore can be seen as an innovation of today's understanding.

3.2   Three Potential Waves of Innovation Development

Kondratjew (1926) has introduced the idea of 50-60 years longing waves describing economic development since the end of the 18th century. For him each cycle or wave begins with an introduction of new technology, which was in the beginning the appliance of the steam engine type by James Watt in the 1774 (Deutsches Museum 2000). Schumpeter took the concept by Kondratjew further and actually coined the term "Kondratjew wave" in his work of the Business Cycles (Schumpeter 1939, pp.173–174). Schumpeter also believed in the creating power of destruction and innovation was the mean of creation. Neither Schumpeter nor Kondratjew described in depth, however, the differences between these revolutionary changes and their different origins and motivations. A question would be, why did someone innovate and who was influencing him or her?

Mankind needed new products and achievements, in order to survive and adapt to the hard and changing life conditions. It was probably curiosity and unsatisfied circumstances that on the other hand have pushed peoples to look for new places to live and thereby redoing the circle of adapting and surviving.

3.2.1 First Wave of Innovation

Till the age of Mercantilism and later Enlightenment however, there were two prominent sources of power – the aspirations of the church and the monarchy – that have probably influenced people and society the most to move beyond this form of adaption and surviving. Only a few exceptions can be found, for example on the Greek Islands and in some parts of India or China, where philosophers and scholars have thought and developed ideas and concepts, without being concerned about existence, religion and politics (Babinotis n.d.). Further developments in those times, especially in technical or engineering areas, were mostly in the name, sponsored or inspired by religious leaders or a monarch – whether it was the construction of churches or castles, the design of ships or the invention of the stirrup. Thus, innovation went to a similar cycle as art, which was also primarily used to serve those two powers, painting the monarchs and their battles or telling religious stories (Krauße 2005, pp.52–53). Therefore, early innovation also has to been seen in this context. This phase could be defined as the first wave of innovation – starting the analysis at the end of the ancient world, leaving out the discoveries and inventions in ancient Greece, Egypt or China. The introduction of the stirrup (see also 3.1) in Europe in the middle age, supporting the raise of feudalism from the 6th century on, could then be seen as the first innovation and thus marking the beginning of this innovation wave (Encyclopædia Britannica 2014).

3.2.2 Second Wave of Innovation

With the second wave of innovation during Mercantilism, new drivers and characteristics were added to the innovation process. In the beginning, it was wealth creation and stability for individual economies that led to state monopolies and the introduction of patents and business protection (Rothbard 2006, p.213ff) – which are today very typical business criteria. But later private individuals also profited from these changes. Although still being closely connected with the monarchs and the church, it was the raise of merchant dynasties in Britain, the Netherlands and later in Italy, as well as the banks that opened new frontiers for innovation, for themselves as well as for those in interaction with them (Magnusson 2002, p.21ff). New banking systems fueled the rise of financial institutions and businesses (Hoggson 2007, p.63,71ff); international trade and the shipping industry flourished and the printing of books set the ground for mass education. With improved vessels and navigation this international trade further developed into globalization through colonization around the world (About.com 2014). With the improvement and invention of new instruments like the hammer piano in Florence around 1700 and the tuning fork in 1711 (Spiegel 1995) this time was also the beginning of the great musicians that reached their tipping point with Bach and the Baroque Age and from then on influencing music, opera, ballet and the music industry till today. The invention and introduction of the tuning fork for example can be seen as a classic innovation, although its purpose was not a monetary market penetration. It was introduced on the music market and improved the systematic tuning of instruments, which led to more constant and better music composition and replay. This also had an effect on the success of musicians and instrument makers, thus leading to modern understanding of financial success of an innovation.

3.2.3   Third Wave of Innovation

The third wave of innovation then came with the Industrial Revolution, reaching a level, where innovation is more comparable with our modern understanding. It was the time of great inventions and market introduction of the steam engine, the light bulb, or the automobile that connected innovation with business success. These opportunities were no longer only accessible to noblemen but also to the general public and spread over the world, from England to Germany and the US. That is probably the reason why most research and documentation on innovation starts in the early 18th century, with the rise of economics. Adam Smith (Smith 1776, p.10) was somehow describing innovation and the motivation of men to be part of it, without using the term innovation. "Kondratjew waves" therefore have their starting point in this third wave of innovation. It was the time, in which big companies and large corporations were founded and in which individual success created rich families from different backgrounds. It thus was also the beginning of corporate innovation, which has been further developed till today and has created institutions like the Bell Labs or modern think tanks in various companies.

The military importance and its requirement for better weapons, logistics and tactics however, has always been there throughout those waves – till today. And research and development in the military sector as well as the half military, half civilian area of aeronautics leads nowadays to more great innovation, which also influences and pushes civilian innovation further. And also the after effects of wars like civilian destruction or human injuries require innovation to reconstruct and heal. A good example is the research and application of Professor Joachim Kohn (2013), who is helping US veterans to heal and grow new skin or bones through a groundbreaking biochemical process. But also crises from natural catastrophes can inspire inventors and entrepreneurs to come up with new application or products, to help people and/ or seize business opportunities. Drones – rather known in a military use case – are getting developed to deliver necessary supplies, where other forms of transportation fail, for example in areas with destroyed infrastructure after an earthquake (RWTH Aachen 2013). Maybe these forms of innovation are already an example for a forth circle of innovation, in which innovation is implemented by entrepreneurs, who mean well for society, and thus are not or are not primarily driven by business purpose. Nowadays, they are often reflected as "social entrepreneurs" (Ashoka 2013).

	First Wave	Second Wave	Third Wave	Next wave?
Time	6th - 16th cent.	16th - 18th cent.	18th cent. - today	21st century -
Driver	Military & church	Money & trade	Industrialization	Social responsibility? Crowd Intelligence? Open innovation?
Charac.	Power, military enforcement, construction	Patents, monopoly, economic wealth, banks, merchandise dynasties, navigation, print	Technology, inventions, corporations, individuals success, growth	Social entrepreneurship, foundations, corporate social responsibility crowd activities, free IP, social media

Table 1: Three Waves of Innovation after the Ancient World

3.3 Further Defining Innovation

Schumpeter has shaped the word innovation and its meaning in the 20th century for generations of researchers and business people. His most important distinction is in this context separating the invention process from the implementation part and connection the implementation of the invention, and thus transformation into an innovation to the entrepreneur (Schumpeter 1911). Only if the two come together, we can talk about an innovation. A researcher, who has invented a new product and even filed a patent for it, has only done the first part of the equation. If neither the entrepreneur, nor anyone else brings the product into a market, it remains an invention. What appears to be simple, is often in modern research misinterpreted.

3.3.1 Type of Innovation

Many scientists even believe, that once they have published their new achievement, they have done an innovation. On the other hand, a businessperson, who only copies an existing product or business model for example, has not caused necessarily an innovation, just because he introduced into a market. If the market is already acquainted with the product, it is not an innovation. However, if it is introduced on a market, where it has not been before, one can call it a **market innovation** (Gaul 2002). If the product, for example, is introduced in the US market for the first time in the world, it is called **world innovation** or absolute innovation (Pepels 2004). If then someone else copies it and introduces it to the German market, it is a market innovation in Germany. Pepels does not see the differentiation in world and regional markets like Gaul, but he defines a so-called **company** or **relative innovation**, which is new to a company but not to the market (Pepels 2004, p.447).

The local market innovations are a modern and very common scenario, especially in the Internet business, where a lot of businesses are being copied from the US and introduced to a new market. Examples are Ricardo as a copy of ebay, or CityDeal as copy of Groupon or NineFlats as a copy Airbnb (also compare chapter 5.13 on internet copycats).

Focussing on the world innovation always requires some sort of creativity, which is seen by many as the seed of all innovation (Amabile 1996, p.1155). And here, creativity designates the ability to create, not imitate, and transcend the known ideas and models and being original. Some will therefore argue, that those doing a so-called copycat are more or less managers that take a business to another country. They lack the creative part and the initial risk taking, to see how customers react to the product or service. This is expressed by Steward, Jr. and others (Steward 1998, p.195), who say that innovation and creativity are inherent conditions of entrepreneurs and separate them from managers.

Being a successful innovative company, therefore, requires the ability to nurture creativity and transfer its results to a business concept, which can then be implemented. Employees need motivation and support in supplying them with resources, such as time, tools, training or facilities as well as the freedom and autonomy in being creative in their daily work life. Feeding the employees with interesting and challenging work can be inspiring, and thus part of an innovation process (Amabile 1996, p.1156). It depends very much on the founders or managing directors and their abilities to create a successful organizational culture, which supports the whole innovation cycle.

The Organization for Economic Cooperation and Development (OECD) has defined innovation in four different ways. It distinguishes between **product-, process,- marketing-** and **organizational innovation** (OECD & Eurostat 2005). For them a product innovation is a good or a service, which is totally new or significantly improved. Improvements can be in technical specifications, components and materials, incorporated software, user friendliness or other functional characteristics (OECD

2003). A typical example for a process innovation would be the introduction of the assembly line production for the T-Model by Henry Ford in 1913 (Batchelor 1994, p.37).

"*A marketing innovation involves a new marketing method involving significant changes in product design or packaging, product placement, product promotion or pricing*" (OECD & Eurostat 2005). Good examples can be found in the Internet, such as affiliate marketing. Companies can offer services or products and promote them on their website, although they neither produce or have them. Once the user decides to have that product or service, he will be linked to the actual owner or seller of the product. The owner or seller of the product rewards the company doing the advertising. This type of marketing innovation was even filed for patent in 1996 and introduced by William J. Tobin, the founder of PC Flowers & Gifts (Habeeb n.d.).

An organizational innovation can, for example, be the Japanese Keiretsu. In this system, the huge multi-company groups in Japan are anchored by cross-shareholding. Japanese companies are more inclined to choose to subcontract with other companies than merge with them. This arrangement works certainly for long-term management, and it has enjoyed outstanding success for the last two decades, because it strengthens the cooperation and mutual interests in each other's success (Gabler n.d.).

3.3.2 Innovation as a Must for Survival

Innovation is not only an issue for a company's R&D department, because it needs among others a close cooperation between marketing and sales, in order to find the right customer taste and generate sales. Today's business dynamics require companies to come up with new products, business models and service innovations in order to compete and stay profitable in their markets. If they are not doing so, it is again Schumpeter, who says that competition will literally destroy the company (Schumpeter 1947, p.163ff). Every company has a natural life cycle of their business existence, which can only in the long run be extended through innovation (Würth 2003). The ability to innovate is, therefore, a difficult but essential asset of successful company. No wonder that in a ranking by the Boston Consulting Group, for example, Apple and Google were leading the list of the world's most innovative companies 2006, which are known to foster innovation within their company (Paris et al. 2006). Gary Hamel (2007, pp.51–53) from London Business School believes that every single employee of the company could and should participate in the innovation process. In his book "The future of management" he says that the mission of an organization is to utilize the creativity of its employees. They must be given the tools and the time to use this hidden gift (p.52).

3.3.3   Origin of Innovation – From Ideas to Inventions

As innovation starts with an idea or invention, it is interesting to see, where the ideas for the inventions are coming from. Ripsas (1997, p.138) has come up with a spread showing the distribution of sources of ideas:

Source of idea	%
Employment in the same field of business	43%
Hobby and activities in the leisure time	16%
Effort to make something better as you have	15%
Realization of an unused market niche	11%
Systematic search for a good idea	7%
No feasible explanation	8%

Table 2: Sources of Ideas for Inventions (Ripsas, Sven 1997, p.138)

It is interesting to see that approximately half of all ideas have their origin in the daily life of each employee. Schumpeter advances the view that the entrepreneur, driven by profit, which is related to innovation, is going to accomplish the new developments and, in doing so, he destroys established balances on the markets, and thus ignites an on-going change and competition for improvement (Schumpeter 1947, p.138ff,143ff).

Schumpeter also emphasizes that the competition among companies shows also another source of innovation, not only coming from price and quality pressure but especially from new technologies, methods and organizational changes. This change and competition can happen through new combinations in the following fields, which also entrepreneur and professor at the University of Karlsruhe, Rheinhold Würth, describes in his book "Entrepreneurship in Deutschland" (Würth 2001, p.37ff):

- The manufacturing of new products and product quality
- The introduction of new production methods
- The exploitation of new markets
- The capture of new sources of supply for resources and intermediate goods
- The reconstruction of sectors of industry.

Würth (2001, p.140) mentions that a successful innovation should fit the preferences of the customers, additionally the right form of marketing should have been chosen, and the innovation should stand out from other competitors.

3.3.4   Numbers and Monetization of Patents and Innovation

As ideas and inventions are a base for innovation, their numbers listed at the statistical national departments give some interesting insights. For Germany and the United States of America for example the number of 581.67 for Germany and 800.17 patents for the US per 1 Million inhabitants can be found (Destatis 2008). Given these numbers, Germany and the US have a similar base for innovation; Bloomberg lists the US as number 1 and Germany as number 3 in the 2013 ranking of the most innovative countries – judging on a mix of patents, high-tech density, productivity, tertiary efficiency and some more (Bloomberg 2013). The US however, outperforms Germany, when it comes to market capitalization by far: 1.486 trillion USD for Germany stand against 18.668 trillion USD in the year 2013 (Worldbank 2013). It indicates that the US must be very effective when it comes to the monetization of its patents, business ideas and development of its companies, when believing in Schumpeter's understanding of capitalism (Schumpeter 1947). And this exploitation and implementation in the market is for Schumpeter the major role of entrepreneurs.

Nevertheless, these numbers must be interpreted with caution, because Switzerland for example accounted in 2013 for a market capitalization of over one trillion USD, which is far un-proportionally to its size. Maybe it also shows how the financial sector has gained leverage. Beside Switzerland with its immense financial industry and global importance, also the UK depends today on the financial sector and outperforms Germany with a market capitalization of over three trillion USD although their GDP is lower than Germany's (Hubbard & Dudley 2004). A financial example for this can be seen in 2004 when the London Stock Exchange reckoned 285 IPO while the German *Deutsche Börse* had only 20. This is an evidence of financial success and countries being powerful without leading in innovation. Dudley, a chief banker of Goldman Sachs, and Hubbard, former dean of Columbia Business School, however, had the thesis that these developed capital markets lead to several benefits that exceed the financial side, but also reach productivity and employment opportunities. This positive influence on the other economic areas might be doubtful, especially, when seeing the collapse of the financial system in 2008/2009; tearing down also the classic industries that had been well and healthy performing. However, this relation exceeds by far the scope of this thesis; it only challenges the modern understanding and influence of innovation based on the thoughts of Schumpeter. It also leads to the next section about entrepreneurship and the question, who bears the risks and carries out the change for innovation.

3.4 Defining Entrepreneurship & Entrepreneur

The word entrepreneur has French roots and was shaped in the works of Jean-Baptiste Say (1821), who defined an entrepreneur as *"someone, who undertakes an enterprise, especially a contractor, acting as the intermediary between capital and labor"*. He also added, that this person basically takes on a risk in a pursuit of profit by shifting resources out of areas of low productivity to the areas of higher productivity.

The etymological origin comes from *entreprendre* - to undertake (Frederick & Kuratko 2010, p.4) and was already in use in the 17th century, in which the *entrepreneur* was an individual appointed to undertake a particular commercial project (Dellabarca 2002, p.12). That is similar to a dictionary definition, which says, that an entrepreneur is "one who organizes, manages, and assumes the risks of a business or enterprise" (Merriam-Webster 2010a).

Entrepreneurship can then be seen as the activity, which an entrepreneur is performing. However, there are many different definitions of entrepreneurship, not only in academia, but also in countries like the US and Germany.

In the United States entrepreneurship is the profession of an entrepreneur, whereas in Germany the direct translation to *Unternehmertum* often stands for the behavior of all business owners, stakeholders or higher management of a company to act entrepreneurial; for founding companies and changing as well as implementing new aspects (Gründerszene 2013) and thus be an *Unternehmer*. However, in Germany often business owners, especially those who have inherited a company, might still be regarded as an *Unternehmer* although they rather act like a manager, who owns the company or share of it.

3.4.1   Differentiation of Unternehmer, Entrepreneurs and Intrapreneurs

The distinction between *Unternehmer* and entrepreneur fits also in the context of various founder forms, described by Szyperski & Nathusius (1999, p.27). Someone for example could become an *Unternehmer* by inheriting the company from the family, without ever having founded or created something new. It would be a derivative self-employment. Or in many modern Internet scenarios, successful entrepreneurs or investors have an idea, for which they find a sort of manager, to build up a company. This would mean to be not self-employed though working in a genuine or derivative company. Even if that manager receives equity of the company and can shape things according to his or her belief, the question remains whether it is appropriate to call him or her an entrepreneur? The same question can be raised for a new wave in existing companies, who look for entrepreneurs within their own organization. Some call them *intrapreneurs*, what at least signals that these individuals are different from the entrepreneurs that have been described above, but still have some things in common. Pinchot III and Pinchot (1978) are said to have mentioned the term intrapreneur for the first time in 1978. For Szyperski & Nathusius these intrapreneurs would either create a spin-off with a genuine idea or just take an existing business part into a new form of independency through a new company foundation. In both cases the startup would still belong at least in parts and have ties to the original company. However, some companies already use the term intrapreneur for employees, who stand up for their own ideas and take some risk and responsibility within the corporation. In such form, not even a foundation of a company has to be considered (Jong & Wennekers 2008).

	Derivative	Genuine
Not self-employed	Transformation founding	Spin-off (with ties)
Self-employed	Overtaking a company/ business	Company foundation

Table 3: Various Founder Types Source: Szyperski & Nathusius (1999, p.27).

Schumpeter (1911, p.78) added a different component, because he saw entrepreneurship as innovating something totally new or making something new by combining existing ideas. In his view, entrepreneurship was not a profession or a lasting condition of a person, because the person loses this character when he settles down to run the business, which is more similar to management than to entrepreneurship.

3.4.2   Entrepreneurs vs. Financial Risk

Remarkable is that Schumpeter did not see the entrepreneur as the risk bearer of the venture (Schumpeter 1911, p.137). He claimed that the investor – or as he called it the capitalist or financier – carried the risk of the venture. The moment entrepreneurs invest in their own startups, they also become investors, and what they risk is their financial investment. This belief is even more interesting, when considering the real beginning of venture capital, which has started slowly after World War II (compare chapter 3.11) and years after his works. Schumpeter though did not reckon the personal risk of time and failure as part of the risk bearing. In the section above it has already been mentioned that the finance sector might take over parts of the importance and leadership role from innovation. But considering Schumpeter's separation of the risk taker being the financier and the entrepreneur being the "doer", one could define modern investment bankers and hedge fund managers as entrepreneurs trying out new concepts, financial products and methods, while their shareholders are the financiers carrying the risk. This calls for further analysis of entrepreneurship and definitions, being bewildered by following through in some parts on Schumpeter's understanding and separation.

Leibenstein (1968) has altered the definition of Schumpeter stating that entrepreneurship is about gap filling of production functions, which arise from market imperfections. It is about input completion, which is about bringing the necessary inputs together, to achieve the desired output. It is also about crisis management as well as leadership and eventually about putting everything together, formally in creating a firm.

3.4.3   Further Definitions of Entrepreneurs

There are also much more simple definitions like *"entrepreneurship is a role that individuals undertake to create organizations"* (Gartner 1988). Or one of the much cited and internationally accepted entrepreneurship expert Timmons, who summarized it in one of his works as the following: *"Entrepreneurship is defined as any attempt to create new business"* (Timmons et al. 2004, p.2).

The German KfW associates entrepreneurs directly with a startup, in which the founders want to grow and innovate, in contrast to people that simply want to create their own job without innovation and the intention for creating more jobs (KfW-Research 2012, p.5).

Especially in the globalized business this attempt should not stop at the border of a country. Thus, it is good to add the description of McDougal et al (2003) about international entrepreneurs, who say that its innovative, pro-active and risk-seeking behavior crosses national borders and is intended to create value in organizations around the globe.

It is interesting to go back once again to Schumpeter (1911, p.75), who believes that entrepreneurs do not find or create new possibilities, but they have the duty to bring them to life, which brings some complications with it. What about many entrepreneurs like Bill Gates or Richard Branson, who have had their own idea and implemented them and became entrepreneurs? Schumpeter throughout several of his works seemed to be not too stringent on his statements, because he also mentions that entrepreneurs used to be investors themselves or could be the inventors, when later neglecting the statements. Nevertheless, with his definitions and understanding of the different roles and stakeholders he set the ground for an academic distinction, which needs to be continued and enhanced.

Harvard Business School started to use the working definition of entrepreneurship as *"the pursuit of opportunity beyond the resources that you currently control"* (Stevenson 1983), which brought in a few more important aspects such as the opportunity side or control of resources. It very much moves entrepreneurship into the light of uncertainty and daring venture. However, Stevenson also mentions that he wanted to point out with his definition that entrepreneurs are not going for the risk, but rather see the opportunity, which is in a risky environment due to the uncertainty and lacking resources. He even says that the entrepreneurs he knew at that time rather tended to lay off the risk to others, like investors or partners (Schurenberg 2012). Timmons did some fine-tuning on this definition declaring that *"Entrepreneurship is the process of creating or seizing an opportunity and pursuing it regardless of the resources currently controlled"*

(Timmons & Spinelli 1994). This definition is today also used and accepted on the other side of the US, at Stanford.

When we think of what the word entrepreneur means to the layman, one may come up with an image of a person who has his or her own business and strives for independency in pursuit of ones own ideas. Another view is that an entrepreneur is one, who does not fit into any job profile, and thus needs to create a job for her or himself. For many, entrepreneurship also involves risk taking and living with uncertainty. And eventually one thinks of this kind of stubborn decathlete type of person, who wants and maybe needs to do everything on his- or her own – to attain the maximum 1,000 points in as many of the ten disciplines as possible.

In the end of an interview with George Gendron from INC. Peter Drucker said that there is there is only one definition: "An entrepreneur is someone who gets something *new* done." (Inc. 1996)

3.4.4 Social Entrepreneurship

Throughout the last years the term social entrepreneur has also been used more often and extended the entrepreneur definition, referring to an entrepreneur that not only or primarily strives for a financial success but who wants to achieve something positive and social for society (Ashoka 2013). The main intention from the social entrepreneur should be helping underprivileged people or solving a particular problem of society and not making predominantly money. However, social entrepreneurship should not be mixed up with a non-profit engagement. Peter Drucker defined social entrepreneurship in an interview as the change of the performance capacity of society (Inc. 1996). He further says that those entrepreneurs after having had some financial success are looking for a parallel career to give back to society – not only by giving money to charity. By doing a social venture they do not have to change their job, they just have to focus differently.

With this quote Drucker not only defines social entrepreneurship but indirectly leaves space for an interpretation how entrepreneurs in general distinguish themselves from people, who for example just give money to change something. An entrepreneur wants to do more and bring about a change through their jobs and actions.

3.5 First Definitions of the Entrepreneurial Ecosystem

An ecosystem is a word originally coming from biology and is a special form of a biological system. The term system is derived from the Latin word systēma and the Greek word σύστημα systēma meaning whole compounded of several parts or members and can today be described as a set of interacting or interdependent components forming an integrated whole (Merriam-Webster 2011). Systems share characteristics like having components with a certain structure, behavior between each other and interconnections. Their functions occur in physics, biology, politics, economics or social life.

According to Mader *(2011, p.770)* *"An Ecosystem is a community interacting with the physical environment. An ecosystem possesses both living (biotic) and nonliving (abiotic) components"*. The term was first introduced by the British plant ecologist Arthur Tansley in 1935, who started a vivid discussion on the differentiation of the terms, definitions and components of the ecosystem (Tansley 1935, p.297). Although he described the ecosystem on a very abstract level, he defined important elements of the ecosystem in his work "The use and abuse of vegetational terms and concepts", which are today still valid such as the matter of time and size or dependencies and interaction. Already in the 19th century limnologists like Forbes and Thienemann already used parts of today's definition for ecosystems in the description of organic systems for lakes, in which its components were behaving under a certain overall regulating law and interacting in a defined environment (Golley 1993, pp.44–66). Crucial criteria for defining an ecosystem today are (Jax 2004)

- Independent elements interaction with each other
- Defined environment (open or closed)
- Laws and regulations for the interaction
- Time dimension reflecting on the relation of the elements
- Purpose of existence

Moore (1993) is said to be one of the first to have transformed the ecosystem, its elements and laws into economics with the idea of a community of organizations and individuals interacting in a business context.

The term is, however, relatively new to the venture and entrepreneur community. Spilling (1996) took the thought of Moore further, with his theory of an entrepreneurial system. He describes the complex and diverse interaction of the participants, their roles and the environmental factors surrounding them in a certain closed geography or community. A little earlier, Van de Ven (1993) had already connected the participants of this system with the venture creation process, and thus added an important piece to the puzzle of for Entrepreneurial Ecosystem.

Today, ecosystems are a common form used in various economic and organizational descriptions e.g. the iPad Ecosystem, the Cloud Ecosystem, the Mobile Banking Ecosystem etc. (Darlin 2006; Rothschild 1995).

3.5.1   Boulder Entrepreneurial Ecosystem

During the first years for this study between 2009 and 2011, there still was no general agreed upon definition for an Entrepreneurial Ecosystem, and the usage of the term was still vague. Neck et al (Neck et al. 2004) had tried to design a complete and holistic Entrepreneurial Ecosystem with a model based on study results of the Boulder and some other venture incubation communities. Yet, they mainly added cultural aspects to the classic financial and business aspects and without not really completing the holistic approach, especially they did not explain the interaction and regulations of the system as a whole but rather stuck to the individual elements and their role. In addition, other influencing factors like university or informal support were only vaguely described and connected, thus making the model incomplete.

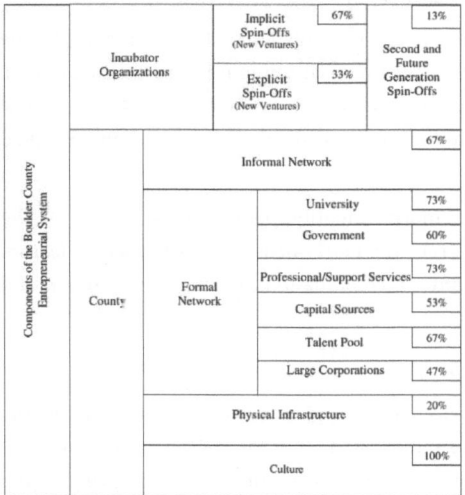

Figure 1: An Entrepreneurial System of New Venture Creation (Neck et al, 2004)

3.5.2   Developments of Further Entrepreneurial Ecosystems

It was professor Isenberg (2011) from Babson College and members of The World Economic Forum in 2011 and 2013 that further shaped, defined and virtualized Entrepreneurial Ecosystems (World Economic Forum & Booz & Company 2011; The World Economic Forum 2013). When talking to investors in the US, the first appearances, however, can be traced back to the Stanford and MIT Entrepreneurial Ecosystem, which did it is not describe or state, what exactly is meant by it or of which components exists – compare the Stanford Entrepreneurship Network. Generally speaking, one can assume that the term was introduced and used, in order to better describe and understand, what influencing factors of entrepreneurship succeed. Since Stanford and MIT are worldwide role models for entrepreneurship, one must particularly look at the surrounding circumstances there. Ecosystem, as a term in business, is already in use for describing other economic environments and frameworks. Thus, it must have occurred that diverse groups like investors, researchers or the government officials have been starting to use the term also for entrepreneurship. The Biology definition from Mader can be perfectly transformed to an entrepreneur's situation. The entrepreneur and his company makes up the community, which is interacting with the market environment and its living components, like employees, family, investors or customers, as well as the non-living components such as the economy, laws or products connected with business. Like all systems, it has boundaries and its components can only interact with each other within to certain laws.

Innovation and entrepreneurship have been recognized by governments to be important for their economy and by parts of the financial community to be also lucrative investments. Therefore, we can see a global trend to scrutinize the Entrepreneurial Ecosystems around the world, in order to find patterns of success, which can be copied or altered.

Between 2009 and 2013, more and more scientific research occurred in the Entrepreneurial Ecosystem. Before, research was more focusing on the success factors of companies and entrepreneurs. Thereafter, a combination of skills, environment and other issues were included in the ecosystem.

3.5.3   Entrepreneurial Ecosystem by Isenberg

Daniel Isenberg is one of the first researchers to give a more detailed view on the Entrepreneurial Ecosystem in an article published in the Forbes magazine. In it, he describes what the characters for the biotic and abiotic components in entrepreneurship are (Isenberg, 2011).

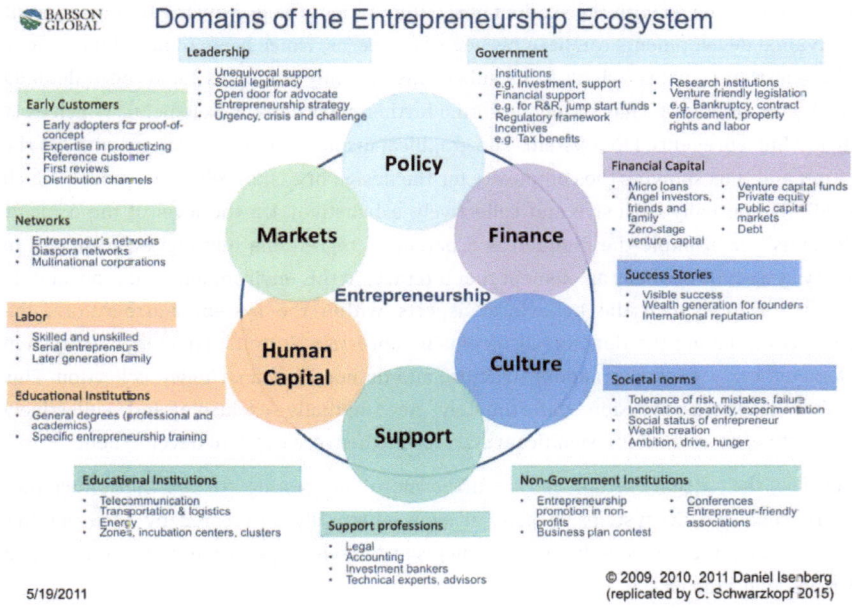

Figure 2: 6 Domains of Entrepreneurship (Isenberg, 2011)

He points out that every Entrepreneurial Ecosystem consists of six major domains:

- Policy: From Leadership to government
- Finance: Financial capital such Business Angels or Venture Capital
- Culture: From social stories to societal norms
- Supports: From infrastructure, to support professions and non-governmental institutions
- Human capital: From labor to educational institutions
- Markets: From networks to customers

These general domain components exist together of a dozen subgroups and its many specific elements. As seen above in the figure, the domain 'Support' consists of 3 subgroups – 'Infrastructure', 'Support professions' and 'Non-governmental institutions'. The support by family and friends – a very common and important element for each founder, however is only present in the financial domain, limiting this element to pure financials. Other missing aspects can be observed for the subgroups 'Network' and 'Early Customers' of the main domain 'Market', in which elements like competitors or technology maturity are neither mentioned as a subgroup, nor are they part of the proposed subgroups. Perhaps that is why Isenberg further postulates that every entrepreneurial – as every natural – ecosystem is unique. Nevertheless, he still believes one can describe all with the six domains, without mentioning, how the totally different individual developments can fit in his general patterns, which seem to have blank spots. The question remains, whether his subgroups are only examples for a more detailed cluster, or whether he has simply forgotten further subgroups and elements, which exist in any Entrepreneurial Ecosystem. The graphical display in figure 2 rather indicates the latter, and thus showing the complexity for the design of a comprehensive model, which is MECE (mutually exclusive and collectively exhaustive). On the level of the 6 major domains, for example the founder as a person is regrettable missing, even though in every system he or she is an element that interacts in this environment and is absolutely necessary to explain the behavioral aspects within the system. Furthermore, the connection among the domains also remains uncertain. In total, Isenberg's ecosystem elements carry a certain random structure and do not show a stringent collection. This makes Isenberg's version unfortunately not mutually exclusive and collectively exhaustive, however it is a significant start for further research and development.

Besides the actual setup, it is furthermore interesting to examine, whether Entrepreneurial Ecosystems in general evolve naturally or created by each nation's business activities or even the startups themselves. Isenberg's answer to this question in Forbes Magazine was:

"They are usually the result of intelligent evolution, a process that blends the invisible hand of markets and deliberate helping hand of public leadership that is enlightened enough to know when and how to lead as well as let go the grip in order to cultivate and ensure (relative) self-sustainability"(Isenberg, 2011).

3.5.4   Two Entrepreneurial Ecosystems by the World Economic Forum, Stanford University, Ernest Young and Booz & Company

The World Economic Forum (WEF) in cooperation with Booz & Company has pictured an ecosystem with the entrepreneur in the middle and various influencing factors in surrounding circles prior to the ecosystem draft from 2013 (World Economic Forum & Booz & Company 2011, p.12).

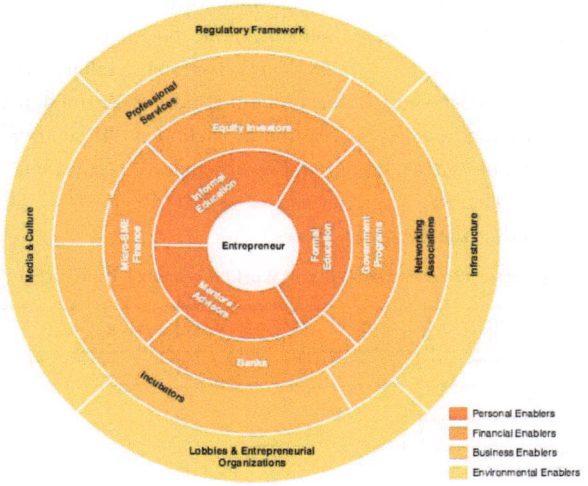

Figure 3: Entrepreneurial Ecosystem by World Economic Forum and Booz & Company in 2011

However, the members of the WEF have not further developed that model till their published a new version in its 2013 report and have not put the elements and relations in a stringent order. The entrepreneur and his or her characteristics have, unfortunately, also not been defined, analyzed and brought in a relation to the other elements. Yet, this circular approach is similar to the author's initial thoughts in the beginning of this thesis, in 2009, and has since then been further developed. It will be a key result of this thesis.

The World Economic Forum designed a new version in 2013, in a corporation with Stanford and Ernest Young and outlined the third and probably most recent sketch of an Entrepreneurial Ecosystem in their 'Report Summary for the Annual Meeting of the New Champions 2013':

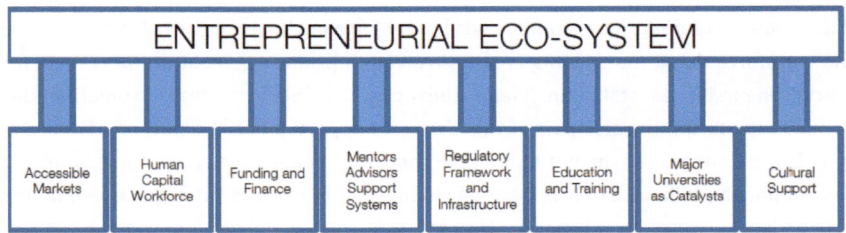

Figure 4: Entrepreneurial Ecosystem (The World Economic Forum 2013)

This version has 8 major domains or pillars as they call them, of which 6 are exactly covering and being similar named like Isenberg's domains. 'Education & Training' and 'Major Universities as Catalysts' are the two others. These two extra pillars are in Isenberg's model combined in the subcategory 'Educational Institutions' of the domain 'Human Capital'. When analyzing the detailed description of each of the 8 pillars and then applying the MECE principle on this version of an Entrepreneurial Ecosystem, the same critique that was valid for Isenberg's version can be stated also for the World Economics Forum's version including the two extra pillars.

Table 4: Components of the Entrepreneurial Eco-System Pillars (World Economic Forum 2013)

COMPONENTS OF ENTREPRENEURIAL ECO-SYSTEM PILLARS

Accessible Markets	Human Capital/Workforce
• Domestic Market – Large Companies as Customers • Domestic Market – Small/Medium Companies as Customers • Domestic Market – Governments as Customers • Foreign Market – Large Companies as Customers • Foreign Market – Small/Medium Companies as Customers • Foreign Market – Governments as Customers	• Management Talent • Technical Talent • Entrepreneurial Company Experience • Outsourcing Availability • Access to Immigrant Workforce
Funding and Finance	**Support System**
• Friends and Family • Angel Investors • Private Equity • Venture Capital • Access to Debt	• Mentors/Advisors • Professional Services • Incubators/Accelerators • Network of Entrepreneurial Peers
Regulatory Framework and Infrastructure	**Education and Training**
• Ease of Starting a Business • Tax Incentives • Business-Friendly Legislation/Policies • Access to Basic Infrastructure (e.g. water, electricity) • Access to Telecommunications/Broadband • Access to Transport	• Available Workforce with Pre-University Education • Available Workforce with University Education • Entrepreneur-Specific Training
Major Universities as Catalysts	**Cultural Support**
• Major Universities Promoting a Culture of Respect for Entrepreneurship • Major Universities Playing a Key Role in Idea-Formation for New Companies • Major Universities Playing a Key Role in Providing Graduates for New Companies	• Tolerance of Risk and Failure • Preference for Self-Employment • Success Stories/Role Models • Research Culture • Positive Image of Entrepreneurship • Celebration of Innovation

'Education & Training', for example, is also part of 'Major University as Catalyst', because the workforce doing the training is also involved in promoting a culture of respect for entrepreneurship as stated in 'Major University as Catalyst' pillar. Promoting this culture supposed to be also part of the 'Cultural Support' pillar. 'Friends and Family' is mentioned as a component, but only in the context of financing and not, for example, in the supporting pillar. Like in Isenberg's version the founder is missing as well as the connection between the elements. Also the state of technology and the competitive environment is not being considered, although it is, for example, one of the key question areas investors discuss with startups. Both versions are also not covering an integration of the state of the economy and the historical backgrounds as influencing factors and components of an Entrepreneurial Ecosystem, which needs to be analyzed, when defining a new version for an Entrepreneurial Ecosystem later in this thesis. Besides the critique and missing elements, both versions have improved the understanding and definition for Entrepreneurial Ecosystems extensively and are a foundation for a further development.

3.6   Startup & Spin-Off

As innovation and entrepreneurship are about creating something new, they are often mentioned in the context of new founded companies, so-called startups. A startup as a term, however, can already be used for a business that is being planned or about to get started, even though there is no legal entity around it. Others refer to a startup only when it means a new venture in the high-tech sector (Botazzi & Da Rin 2002, p.235). However, when analyzing how the term is used in practice, one can see that especially in the US it is used in different industries and describes very young or just beginning companies, from retail to high-tech business. In a study of Cassar (2004) only companies with operating history of less than 2 years were included. They needed to be financed independently and not from a parent company (Cassar 2004, p.270). In the US and German venture scene however, many companies and investors see "startups" as companies that have not reached the profit zone or are still proving their business model. In many cases these startups could be much older than two years. Besides the financial and time aspect, one can often hear companies considering themselves a startup, because of their attitude, their work environment and their philosophy, which they practice.

Very promising new companies are high-growth or entrepreneurial firms that aim to increase their employees' numbers 50 times or more within 5-10 years. These entrepreneurial firms cover only 3-8% of all new startups, but create 70-75% of the net new jobs in the long run (Osnabruegge 2000, p.92), which makes them very appealing for any nation. For the German bank institution KfW, a startup is run by an entrepreneur, who aims to employ several people from the start and is working on a business concept that offers growth, potential and innovation (KfW-Research 2012). These entrepreneurs often start in teams with a co-founder. Founders, that rather start on their own, due to having no alternative or no intention of employing further people, are considered to be self-employed and are not associated with the term startup. They also operate in classic fields with less growth potential. This definition is important in order to separate in the statistics the founders of startups from the self-employed company founders that make up for a majority of the founded companies each year.

A special form of a startup is a "spin-off", which arises originally out of an existing unit like a company or university and takes knowledge and or people out of that unit (FAZ.net - Börsenlexikon 2014). In many cases, the spin-off already has some sort of operating history, and thus also differentiates itself from a regular startup. However, a research-based spin-off for example uses licensed technology from a university or a public research organization that has never been approved or used in the market, and thus could still be considered a startup (OECD 2001, p.7). When it comes to a corporate spin-off, the venture company becomes independent from the parent company and needs to undergo similar issues any startup will encounter (Rüdisüli 2005, p.8). However, the technology might have been approved in parts in the market and the corporate spin-off will receive further advantages like customer base or financial aid in the first place.

3.6.1   Founding Process

The founding process is the time, in which the entrepreneur decides to bring the business idea to life. It is the turn-around from pure thinking, writing and talking to actual doing. At this stage – often considered the seed or startup stage (see next section) – the founder has to stand up and be counted, thus becoming an entrepreneur. Entrepreneurs distinguish themselves at this time from many other people that simple have not the courage to proceed with their business idea. Especially in Germany, due to cultural aspects it is harder for persons to actually take do this step, because they are afraid of cancelling or failing afterwards, not having thought the process and consequences prop   through. It is probably the positive mentality of starting and failing that seems to make it easier for Americans to do the founding step.

This founding process moves the startup to the actual market, gives investors the chance to properly and officially put money into the venture and brings more seriousness and accountability behind the founding team. The entrepreneurs get also in touch with legal aspects, bureaucracy and duties. They can conduct business operations and build up their own company. Staying too long in a secure university environment or business employment, for example, without actually doing this step, and still being able to hide behind scholarships and fixed incomes, carries the potential founder away from becoming an entrepreneur. From the author's experience at the Center for Innovation & Entrepreneurship (CIE) at KIT, many talented students with business ideas did not manage or dare to take this step. Out of over 250 potential startups, at the CIE during 2008-2011, only 20-30% actually moved from the idea stage to founding a legal company (CIE 2012). Besides the fear of failing, it was also pressure from parents to first get experience at an established company and do something reasonable. It was also the peer pressure, not to give up high starting salaries at consulting, banking or industry positions. A phenomenon seen at many elite universities, where a high percentage of students receive well paid financial job offers after graduation.

3.6.2   Company Stages of Startups

A startup needs to become visible to investors and become legally entitled – through legal registration or foundation – to receive money. However, when defining startups nowadays in a broad used context for a company that in general has recently started something new, which is still not established yet or relatively small, a distinction of the business stages must be made. In addition, entrepreneurs often call their company a startup in hope of giving it still a fresh appearance and representing the flexibility and creative power. This observation could lead to a conclusion that startups are also a state of mind from the entrepreneur's side, not wanting to belong to a classic company that probably has lost its novelty. On the other hand, it also offers space and time for the entrepreneur to prove the success of the business.

Finding a useful segmentation for the startups, the view of a venture capitalist (also referred to as VC) is of interest. For a VC, who will be regarded in detail later in this thesis, startups are divided into stages reflecting their state of maturity. Most VCs and business angels consider 3 to 4 important stages, which they, however, sometimes interpret differently (Centre for Strategy & Evaluation Service 2002, p.3; Manger 2011; Deutsche-Startups.de 2012; 1000Ventures.com 2013). The following stages are also very useful for internal planning and other form of financing, and therefore not only relevant for the risk based financing through VCs. The stages are clustered according to different interpretations:

Stage	Description
Pre-Seed	Far before the actual founding of a startup the idea and concept development phase is often described as pre-seed. Some investors even see the stimulation of a potential entrepreneur to think about a business idea as pre-seed. Very seldom money from outside is given to individuals in this phase. It is mostly business plan competitions, events or university programs that foster and fund this stage. In this stage usually family and friends are the financial supporters of teams that are not bootstrapping (financing the operations themselves). Not only the team and concept are in the development phase, but also a business plan is hardly being found, making it very difficult for outsiders to evaluate the idea and potential. Since this phase usually is running under the radar of any investor, because the actual business idea has not been formulated and the entrepreneur is still in a

contemplating mode to find it, many investors see this phase rather as an internal developing time. Many entrepreneurs also carry different ideas with them over months or years and continue to formulate the actual founding story. Up till then, it is usually too far away from approaching any investor. This stage could happen, for example, also during a university seminar and program, without contact to the outside market world. Investors therefore often see the next phase, the seed phase, as the actual beginning of the startup, where they can hear about a concrete business idea (Manger 2011).

Seed/ Startup

The actual founding period is mostly considered as the seed stage. It covers all aspects from business planning to prototyping as well as testing and selling first products. Thus, it also includes some time before a company is actually founded, as well as time after, when first sales could be made. Many investors are looking in this time for the proof of concept, in which the startup has to show not only that it can do the job, but also prove that customers are willing to buy their services or products. Once the product is ready and first sales are made, the startup moves more or less into the early stage phase. This seed or startup phase is the classical time for business angels to invest. The risk is very high, because of a product, which is still in a developing phase, the customer demand has not been proven and the team could not have shown its business skills. Therefore, it is a good ground for government and institutional programs to provide financial support at this stage, taking over some of these risks the market is not willing to bear. In Germany the *EXIST Gründerstipendium* (Founder's scholarship) by the government or the Innovation scholarship granted by states like Baden-Württemberg or NRW are typical examples for this (more details in chapter 6.7.8.1). In the US there is no financial government support for profit-oriented startups (U.S. Small Business Administration 2014b). As soon as the founding has been conducted (investors sometime refer to this time as the startup phase) special bank conditions apply for the startups such as the *KfW Gründerkredit* in Germany with interest rates between 2.5-5% depending mainly on volume and lending period, or the SBA loan programs in the US starting at around 3% depending on the volume, lending period, maturity and negotiations (KfW 2013a; U.S. Small Business Administration 2014b). The German *Bürgschaftsbank* also carries part of the risk for the actual founders, if the company and they personally cannot pay back the credit (compare chapter Financing of Startup Companies). The SBA offers the same for US founders.

Early Stage   After having developed the product or service to be ready for market sales and after a first proof of concept, the startup has to establish its processes and grow the sales to a substantial and steady amount. This early stage is characterized by the effort to make a real business and company from the first customers, personnel and structure. Business angels and some early stage investors are typical for financing this phase. The special banking conditions continue to exist for the startups (up to 5 years after the foundation), as well as further state government scholarship programs, like *Junge Innovatoren* (Young innovators) are available at this stage. The government fund HTGF (High-tech Gründerfonds) is part of this early stage investor group (more details under chapter 6.7.8.1).

Later stage

Expansion/

Growth

When a startup reaches the later stage - many times considered the growth phase - investors believe it has proven the concept and the general demand through steady sales as well as the ability to run at least a small business. The range of sales required for this prove, usually runs between 10K and 100k EUR a month, depending on the industry the startup is in; 50k sales per month for example seen as a good level for million plus investment (Manger 2011). Especially for US startups, the growth phase might already bring several millions in annual sales and the requirement for a multi-million dollar investment in the so-called Series B, C, D etc. (further details in chapter 3.11.5). At this stage classic VC companies take over the financing. In Germany this happens rather later than in the US, positioning the German investors rather at the higher end of the sales expectations, when it comes to first VC investments. One could therefore separate this phase into expansion for the earlier part of the phase and growth for the later part, which represents more the German investing scheme. However, many US startups receive in the first VC financing round more money than a German startup in its final VC round. (further details in the Series A, B, C financing section)

Table 5: Startup Stages

3.7   Three Areas of Innovation

This thesis has the intention to find means for fostering innovation and to strengthen entrepreneurship. Therefore, it is also interesting to examine the origin and driving force of innovation. The author separates the innovation in three areas:

Corporate Innovation	Intrapreneurship	Entrepreneurship
Company driven innovation with a development and introduction of new products, services and methods. The driving forces are different persons – from the CEO, to marketing and R&D itself. This form of innovation stands for the classic behavior and task of any company that wants to sustain in a market for a long time. Projects usually can use many resources and can have high budgets	Company and personal driven innovation, where an individual or a few individuals try to develop and introduce new products, service or methods, because they believe in the necessity and success of it. They either try to bring the ideas into the market within the existing company or found a spin-off. These projects can still profit from many corporate resources and funds.	Personal driven innovation, where individuals follow personal goals and found a company with a concept, service or product, which is new to the market. The entrepreneurs are engaged from the inventing till the implementing part, however, they do not need necessarily have to be the inventors. Usually there are no funds available unless the entrepreneurs have personal savings.
Responsibility is spread on several heads, especially on the executive management. Decision-making is taken away from the innovators, who should implement the invention.	Responsibilities hold by a few individuals – mostly by the intrapreneur, partly by the executive management. Decision-making is partly taken away from innovators.	Responsibility hold by the entrepreneurs, who usually make the decisions, unless they have investors with a strong influence. Entrepreneurs are the innovators.
Mainly driven by shareholder value and tied by compliance rules, thus reducing flexibility, freedom and creativity.	Still driven by shareholder value and limited due to corporate restrictions, however personal merits as well as reputation are	Highest form of freedom, flexibility and creativity.

	coming to the foreground.	
General occurrence in each competitive company with less freedom and decision range for the individual.	Special occurrence, in a competitive and modern corporate environment, leaving more freedom and decisions to a few employees.	Classic occurrence for the entrepreneurs with full freedom of choice and decision among them.

Table 6: 3 Areas of Innovation

The corporate innovation is a form, which can be found in any type of company that is trying to invent and introduce new products to the market. With an increasing responsibility for the acting persons, one could argue that these stakeholders become more entrepreneurial - moving to intrapreneurship or even further to full entrepreneurship. Aside of the financial stakeholders, the general managers or executives, who are involved in the innovation process, have an increasing responsibility, also correlated to an increasing complexity (as shown in the following figure). Generally speaking, the entrepreneur has to manage the whole company and is thus also responsible for the complete innovation process. The inventor or business development manager on the other side is "only" responsible for his or her area and therefore can focus on it. The typical corporate innovation process is mostly targeting the introduction of new products or services. When entire business areas have to be introduced or modified, or when new concepts must be established, it requires personnel with a broader company focus.

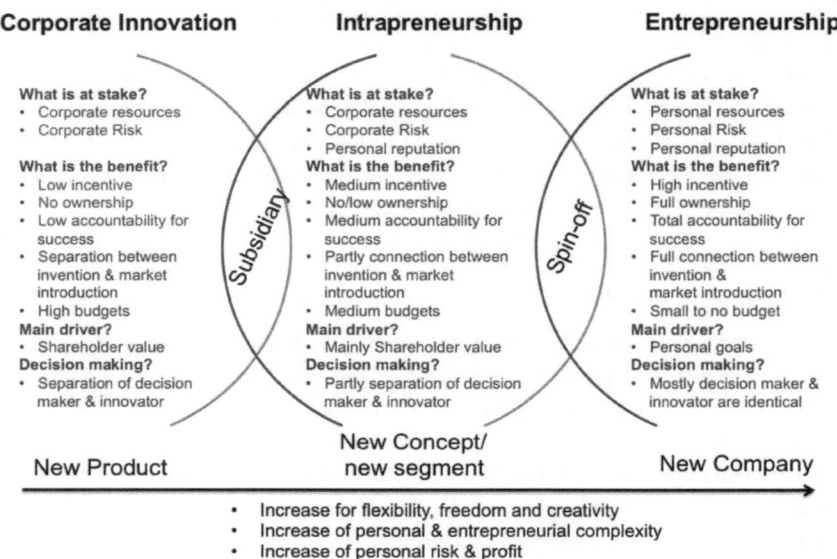

Figure 5: 3 Areas of Innovation - Stakes, Benefits & Complexity

Usually, the inventor is not connected to the market introduction in a corporation, whereas the intra- and entrepreneur are bridging the inventing process to become implementable in the market and push its introduction. The more decision-making is given to the intrapreneurs, the more responsibility and benefit they can take for themselves. In many cases the intrapreneurs act within a given budget and timeframe and certainly as a paid employee of the company. When the innovation process, however, requires to be taken out of the company or when the management is not willing to carry it, the intrapreneur needs to spin the ideas, technology or concepts off the company and found a new external firm. Through these actions, the intrapreneurs move closer to becoming true entrepreneurs. In this case, the mother company usually not only holds equity in the spin-off for its technology but also supports and finances the intrapreneurs in part- or completely. For the author, true entrepreneurship can only be considered, when individuals act independently, without a corporate net securing them in case of failure.

3.8 Open Innovation

When talking about innovation today, the concept of "open innovation" must also in general be mentioned as part of innovation. Chesborough (2003) is said to have coined the term at his time at Harvard Business School, when he specifically examined the changed from "closed innovation" in corporations to an open form engaging externals. Although open innovation is more about the process (Gassmann & Enkel 2006) of how to get to an innovation through, the usage of the power of several external and internal individuals that are not part of one single entity like a company, it is important to be aligned with Schumpeter's concept of dividing invention and innovation – which an entrepreneur carries out. Schumpeter also believed innovation to be more a closed form, because of the exclusive benefits of exclusivity for the entrepreneur. He also wrote that large corporations had been the most successful ones with their innovation. This could be seen in contrast to his belief that entrepreneurs are the ones that carry out innovation (Schumpeter 1947, p.106ff); and only a limited number of entrepreneurs can be found in large corporations. Nevertheless, intrapreneurship on the one hand and open innovation on the other hand seem to solve parts of this dilemma.

Gassman and Enkel (2006) thereby divide open innovation in three areas or processes: an outside in process, in which external knowhow flows into the organization, an inside out process, where knowledge flows in an outside organization (this could be a spin-off) and a mixed form they call coupled process.

On the invention part, open innovation uses several individuals or groups to participate in the developing phases, thus potential IP (Intellectual Property) cannot be really protected. This protection, however, is often a driver for developing new products especially, from a corporate perspective, because it bears huge valuation and market potential (Fisher III & Oberholzer-Gee 2013). On the entrepreneur side, open innovation allows an invention to come easier into existence, because several individuals can also be involved to turn the idea into practice. However, they do not necessarily have to cooperate in the end and can start their own business independently from each other. This principal can work for the corporate,- the Intrapreneurship- as well as for the true entrepreneurship driven innovation. Especially with the use of new technology like social media or online collaboration tools, this innovation driver can play a significant role in the future. On the one hand, more smart people can be involved to solve problems and create new ideas, on the other hands cost and risk can be divided. Modern entrepreneurs and innovation managers can make use of this form depending on their circumstances. The open innovation concept perfectly fits into the Entrepreneurial Ecosystem, because it is a combination and interaction of several players of the system.

3.9 Innovation Management

An innovation is a complex combination of processes and players in various fields from product to organizational innovation. Therefore, innovation management is needed to help to steer through this development (Haussschildt & Salomo 2007, p.30). It consists of individuals, groups as well as the whole organization working individually or collectively (Müller-Prothmann & Dörr 2014, p.118). Traditionally, it is applied in larger mainly product oriented organizations, where numerous decisions have to be made, and running operations and diverse players are involved. Also service companies, as well as startups can use and apply the principles of innovation management (Müller-Prothmann & Dörr 2014, p.48ff).

Using Schumpeter's separation of invention and innovation can also help to further cluster innovation management from two, the creativity and "idea generation" and the perspective of professional implementation perspectives. Both perspectives are necessary in the invention, as well as in the innovation part. Cooper and Kleinschmidt have integrated this line of thought and introduced during the 1990s their second and third generation stage-models, which chronologically order the innovation process from the idea generation to selling of the product or service. There are five connected stages with five "success gates" (Cooper & Kleinschmidt 1990, p.46; Cooper 1996, p.479).

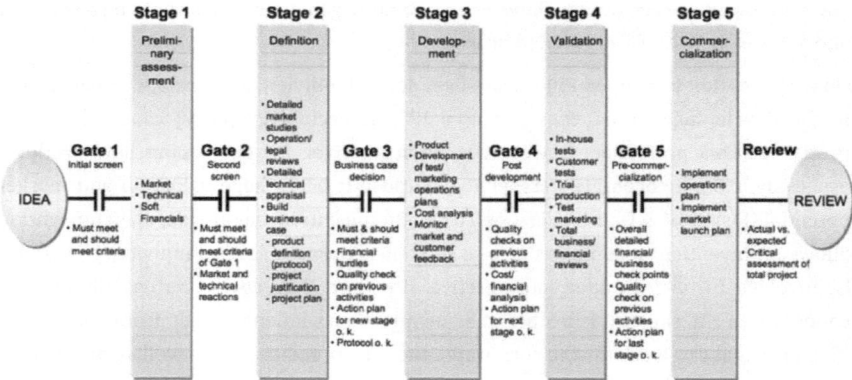

Figure 6: Typical Second-Generation Stage-Gate-Process (Cooper & Kleinschmidt 1990, p. 46)

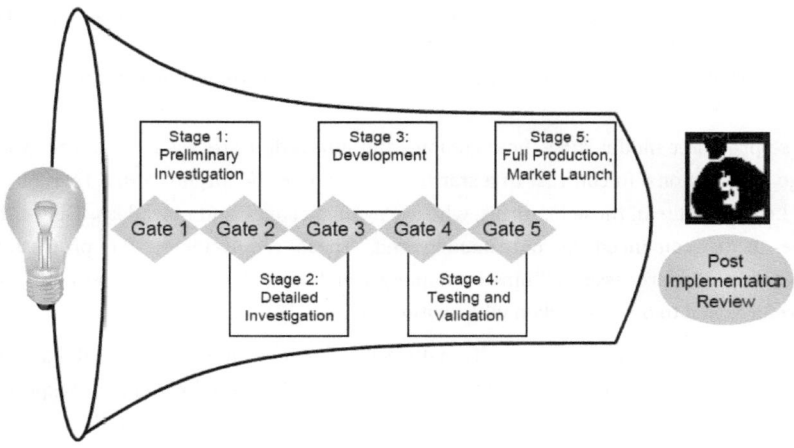

Figure 7: Typical Third-Generation Stage-Gate-Process (Cooper 1996, p. 479)

Six years later, Cooper presents a second model, which still shows a very detailed and a more theoretical approach. In 2002, a third model then came closer to reality, leaving more flexibility in the developing process of the innovation (Verworn & Herstatt 2002, p.7).

In startups the idea creation and creativity process are much more open and flexible (compare Three Areas of Innovation 3.7) than in large established corporations. Therefore, the larger a company becomes, the more it has to invest in its company's processes and organization, in order to create a creative pro innovation culture (Lester & Piore 2004). Creativity techniques like brainstorming and brainwriting can be of help, in order to generate new ideas. Creative blackboards, special mail terminals or dedicated intranet pages are also communication mediums of getting the idea potential from any employment level and employee in the company to the management. It then decides on how to proceed with an idea. Disciplines like organizational management, IT and innovation psychology are crossing at this point and therefore need to interact. Very important in this interaction are the so-called innovation promoters. The German professors Witte and Hausschildt (1999) have developed a model for these promoters clustering in them into four parties:

- Promoters by power, who can engage, motivate and change the innovation process by the power given within the company
- Promoters by expertise, who can influence innovation by their knowledge on the core competence of the company

- Promoters by process, who can influence the innovation process by their organizational and process know-how of the company
- Promoters by relation, who can foster innovation through their in- and external network

These promoter model show, how complex the innovation process is or can be within large corporations. In contrast to a startup, it should not be forgotten that there are or can be opponents in those big firms, who are against change and new ideas. They either have to be convinced or out maneuvered. These requirements for professional structures and processes within a company can hinder the creative processes, and therefore need to be undertaken with caution.

A startup on the other side can by nature develop new ideas easier but lacks the professional implementation, due to its inexperience and limited resources – especially for personnel and money.

3.10 Financing of Startup Companies

The financing of startups is in most cases more complex than those for established companies, because the company has no track record and financial reports. Thus, banks and classic financial instruments have difficulties granting the necessary financial means.

The limited access to capital makes young companies more vulnerable to external shocks and lowers the likelihood of surviving (Saridakis et al. 2008). This part of the chapter will concentrate on the financial needs and problems of the company.

The formation and registration of the company, as well as the reservation of Internet domains are usually the first expenses that arise and cannot be avoided. These could be followed by potential costs for trademarks, design, website support and business cards. Once the company wants to have an office and needs to pay monthly fees for rental, IT, Telco, electricity and so on, it will put financial pressure on the startup team. From observations at the startup incubator CIE at the University of Karlsruhe that teams, which wanted to implement their ideas and inventions, the basics such as furniture, communication, hard- and software, as well as personnel, created monthly cost between 500-1,000€. Those are typical prices for a 3-person team and comparable with fees other startups incubators and shared office locations take (compare costs for rentals at immobilienscout24.de, BetaHouse Berlin or Plug-and Play Silicon Valley in 2010). Most teams also needed help and money in programming and designing their web- and mobile site – as one of their first expenses. All together, the average startup needed around a minimum of 5,000-10,000€ for the first 6 months to get started. An amount too small for most external financial help from Business Angels, but a sum of money already very high for founders coming out of or being in university. Once the team has been founded and needs to finance either each others' costs of living and or having first employees, the costs jump up to 7,000-10,000 € a month.

Of course this does not include any costs for prototyping, material costs and development for highly sophisticated or high-tech companies. These costs could quickly exceed 100,000 € or even much more in the first year for machines, labs and experiments.

So, with this very early need for capital comes the need for one or more financiers. The Modigliani-Miller Theorem (Ebben & Johnson 2006), which says in the case of perfect markets the only thing that the capital structure affects is the division of the cash flows among the investors, does not apply here. Not only because the market is not perfect, but also because of the very high risk of failure and the uncertainty if and how the business will develop. All financiers will ask for relatively high percentage of equity or interest, which the founders often find disturbing, excessive if not outright unfair. The business has not even take off, and already the discussion starts with this equity distribution, leading to awkward decisions and feelings within a startup team.

In most cases, founders, therefore, start investing their own money as well as borrowing from family and friends, making it sometimes easier and quicker to agree on an equity or interest ratio.

For a startup the question of financing is in general about how to obtain it and not about which means or source to choose, due to numerous constraints in obtaining financing from external sources. With family it is relatively simple, because they know the founder or founders and usually believe in him or her and want to lend their support. Starting with friends and going on to other external investors, it becomes more complex, because many more things have to be explained. This leads to constraints, due to information asymmetries between the entrepreneur and the financier (Winborg & Landström 2001, p.851). Information asymmetries may occur when the entrepreneurs possess superior information about the possibilities of their own business, but are not able or willing – because of the fear of possible losing or having stolen trade secrets, which were communicated that to the potential investors. On the other hand, the limited ability of the investor in monitoring and controlling the company make everything more complex (Cassar 2004, p.237).

Nevertheless, the decision about controlling or influencing the source of capital and the advantages and disadvantages of this choice has shown to have an influence even on the later success of the firm (Cassar 2004, p.1).

The new company, which has received funding after qualifying and passing the screening process of the investors, can move forward and even receive further attention and opportunities for more funding. Once the company is up and running, the external investors have a high interest in supporting the company, to secure and multiply his return on investment and hence works together with the startup team.

3.10.1 Personal Financing

It is almost certain that starting a business means investing one's own money. A study of Elston & Audretsch (2010, p.87) found out that 58% of starting entrepreneurs used not only personal savings but also working in a second-job to finance the startup. Often when more money was needed, further financial aid come from close family, friends and other relatives, sometimes referred to as *FFF* – family, friends and fools. About 81% of American new businesses are financed by the founder themselves, while 15% of the new businesses had acquired funding from friends or relatives (U.S. Chamber of Commerce Statistics and Research Center 2005, p.3). A very small portion also receives government grants or has won prizes at business plan or startup competition, which nowadays are not only common in the US, but around the globe – including of course Germany.

Depending on the industry and level of experience of the founders, some startups manage to make early revenues, which can help to finance the company. Especially service-oriented companies can generate this early income. The opposite is the case for product-, software- or Internet startups, which have to invest money and time, before it they sell their products or generate revenues through advertising for example.

A startup that lacks early revenues and funding, can try to grow the company with its own savings and try to run on a very low budget. This type of running a company and the activity of finding creative ways to overcome the financing gap is called *financial bootstrapping*. Bootstrapping includes things like sharing and borrowing equipment, involving low-cost subcontractors to the development process and absorbing resources from both sides of the supply chain (Cassar 2004, pp.235–250). Bootstrapping usually means that the founders fund their activities themselves.

3.10.2 Trade-Off Theory

The first of two competing theories, which try to explain financing decisions in a company, is the static trade-off theory. Described in a paper of Swinnen et al (2005, pp.1–3), its basic assumption is that an optimal financial debt ratio exists, and it maximizes the value of the company. The company will gradually move towards this target. In the optimal point, benefits of issuing more debt are balanced by the costs of issuing this debt. The theory also assumes the existence and access to a perfect capital market, which is in the case of a startup even more unrealistic than for an established firm. It is also very questionable whether startups have the knowledge and the resources to find their optimal financial debt ratio. Furthermore, these thoughts are very hypothetical, because these young companies have many more issues to solve and deal with other more effective levers for success. In addition, startups often adapt or change their business model thus the ideal debt ratio changes as well.

The theory itself concentrates on the pros and cons of issuing debt. The benefits gained with debt are the tax deductibility of the paid interest, known as "tax shield" as well as that it undermines the agency relationship between the manager and the shareholder. Since most startups usually make no profits, tax deductibility is not a priority for them. Nevertheless, there are other negative effects. Due to the costs of financial distress and the agency costs, which are caused by conflicts between shareholders and debtors. Another area, where the trade-off causes major problems for the ambitious and independent entrepreneur, is when he or she has to account to the investor for company decisions and development. The cost of financial distress will occur when the company is unable to honor the payments for its liabilities. The theory sees the firm balancing between the debt tax shields versus the costs of financial distress (Frank & Goyal 2007) and choosing the optimal capital structure respectively.

3.10.3 Pecking Order Theory

The second competing theory to the trade-off theory is the pecking order theory. It describes how and why companies choose a particular financing source and information, in making a particular decision. The main statement is that companies are said to be following a "pecking order" when they prefer internal over external financing and debt to equity if external financing is chosen (Myers & Majluf 1984, pp.219–220). Startups usually have either no or little financing options, and they tend to take what options are available.

Myers and Majluf (1984) came up with the theory about financing preferences of companies between several sources in their paper "Corporate financing and investment decisions when firms have information which investors do not have". The pecking order theory is based on the assumption of transaction costs and information asymmetry, and thus raises the same issues, which the entrepreneur and investor face. It tries to combine the effect of insufficient information and how to increase the assumed risk by the investor. The result is a required rate of return or interest rate, which equals higher cost of capital to the entrepreneur. When overpricing the risk, a result could be that the entrepreneurs or managers come to the same conclusion internal financing is more attractive (Cassar 2004, p.264). It also has the opposite effect; when the company has superior information about a potential negative outlook and this information is not included in the share price, it is willing to give out new shares rather than borrow money. If management decides to make a secondary offering of the share, one should raise questions. This skepticism has also been approved by Ashquit & Mullins (1986, pp.85–88), who have discovered negative stock price reactions, due to announcements about a "follow-on" offering of new or more shares. The goal of the "follow-on" offering is to raise as much capital as possible, thus management is more than happy to offer overpriced shares. In case of undervalued shares, the management might seek another source for financing, thanks to the decision made at the top. For startups this is very interesting, because in many financing rounds teams should try to raise more capital than planned by their calculations. This is because of the uncertainty of the developments and because the bad conditions startup receive, when they are running out of cash earlier than expected. When such financial pressure arises, investors tend to get an over proportional shares for more "fast money". In addition, this extra needed money, will cost the startup team significant more time and effort, thus slowing down or even going under.

Another way of explaining the Pecking Order Theory is by connecting it with transaction costs. Which means that the management team has to compare the costs of using operating income, external debt financing and the issuance of new equity, before deciding in each case on the ideal and suitable financing form (Baskin 1989, pp.26–35). The lacking of constant and high revenues in a startup usually makes this aspect of the theory, however, irrelevant for many early stage startups. The more mature a company becomes or the closer the connection between sales orders and expenses are, the sooner a startup will consider, compare and make a pecking order decision for the financing. If a startup for example receives a sales order of 50k units worth 400k €, and it needs to know the extra production costs for this order to be 150k €. Furthermore the startup needs to compare the interest rate they need to pay for a bank loan with the equity they would give up when finding an investor. This pecking order becomes however very complicated, when on the one, hand there is a risk of an order cancellation and, on the other hand, it usually takes a considerable amount of time for startups to raise money in exchange for equity. In addition, startups usually have a bad credit rating, making it very expensive or impossible to borrow money from a bank.

Considering these factors, the managing team forms an internal pecking order to their preferences of financing the company, when it has the necessary knowledge and the options available.

3.10.4 Debt Security – Real and Personal Property

Borrowing money from classic sources such as a bank is also important for companies, as well as startups. A big advantage of bank debt is that the owner does not have to give away equity, although control might be attached to the company, in form of covenants, in the bank loan contract (Berger & Udell 2005, p.301). If the necessary amount is small, it might be an advantage since less information on the financial data is required. However, it requires in Germany and the US usually a guarantee of the company's owner. The negative part of bank loans is the pre-set repayments of the debt and the interests, which can become a problem, especially if the company does not perform as planned or reaches its financial goals.

Debt represents 52% of the capital structure of young US companies, which are less than two years old (Berger & Udell 2005, p.299). The proportion is about the same in larger corporations. Debt represents still a third of the capital structure just before the IPO in American businesses, and in Britain 66% of the SMEs reported, in 2002, having relied on borrowing from a bank (Burke & Hanley 2006). It is also the most common way of external financing (Meza & Southey 1996, p.377). The loan conditions normally include providing personal property collateral to cover the loan. Internal or inside personal property collateral include some firm-owned securities like accounts receivable, equipment or inventory. This means that secured creditors receive a priority to some assets if the firm is liquidated.

On the other hand, *external collateral* means to include company real property or the right to some property of the firm's owner. Borrowers usually use their house, as a security for the loan. In a young founder's case, this house is often owned by their family (Hanley & Girma 2006, p.353). As a result, the contract conditions, the control rights of the collateral may shift to the creditor. In this case, generally the home cannot be sold without an approval of the creditor. Nevertheless, this should not be mixed up with a *personal security or guarantee,* which means that the owner becomes liable for the loan with his total personal property and must pay, if the debtor defaults. In those cases, a personal guarantee will not transfer the creditor any control rights over the property of the debtor before the liquidation of the company. The amount and the quality of a collateral effects the interests the entrepreneur has to pay, or the amount of financing he or she will be offered (Burke & Hanley 2006). It also works as a strong incentive for the entrepreneur to repay the loan (Berger & Udell 2005, pp.305–306).

In the case of young, innovative or high-risk companies, the banks are normally not ready to bear the risks involved (Eulenberger 2008, pp.69–70). Actually, a study cf Freel (2006, pp.23–25) found out that the most innovative companies have less success in loan markets than their less innovative peers. Banks lack the knowledge to review and to evaluate the firms and thus are not prepared well enough for risk taking in the startup sector. Without real or personal property collateral, they have no mean and interest of helping these companies. As a result, short-term maturities are preferred over long-term debt. Consequently, it is hard to measure the average interest rate which banks give a startup. That is why in Germany the government controlled bank IKB used to fill in the gap, with a fixed credit line of up to 50,000 € and 5,3 % interest rate (together with partners like the KfW) and the KfW directly still offers rates below 3,5%, however after an KfW specific application process (KfW 2013a).

3.10.5 Equity

Due to the constraints in getting financing from banks, startups are rather looking to seek financing from the Private Equity market. This means that the owners are selling shares and parts of their control to outside investors. Thus, the entrepreneurs share the risk of failure with other shareholders and do not have to provide personal collateral. Even though the risk of loss is high, there is a possibility to high returns of the investment, much higher than on the public capital markets. The actors of the Private Equity market will be presented more closely in the following section on risk capital.

3.11 Financing with Risk Capital

Risk capital stands for all types of investments that bear the risk of total loss and that receive equity of a company instead of guarantees, securities and fixed interests. In practice one uses nowadays mainly the term venture capital (VC) instead of risk capital, although VC has multiple meanings from the complete abstract term to a specific subgroup. VC is used for a person that invests risk capital in startups; it further stands for the general group and money of highly risk investments and is also used for the venture capital funds themselves. The main objective of VC is a capital gain accompanied with dividend yield (Wright 1998, p.1).

Venture capital filled-in the financing gap of startups for the first time in the United States after World War II. VC by definition is in most cases coming from private money given to the financing of a company instead of a bank credit, and thereby becoming the private equity of a company. VC is often reckoned to be a part of private equity (PE) business. Private equity is an asset class and more general term for equity capital provided to non-listed companies. PE also includes besides VC the investing in later stages and established companies, such as management buy-outs and buy-ins (Forbes 2012). A little misleading in practice is that one speaks mainly about PE business or money, when mature companies get extra finance, whereas for investing in new or early growth companies one uses the more distinct term VC as a subgroup of PE (EVCA 2010).

In today's VC world, there is a wide investment spectrum, from a startup, which has not even been founded, up to a company that already makes millions in revenues. The size and origin of the money, as well as the background of those giving the money, usually determine the investment scheme. In general there are three types of VC, although one has to say that even these three types are not mutually exclusive and stringent: business angels, private investors and venture funds.

3.11.1 Business Angels (BA)

The term business angel (BA) was used in England for the first time to designate those rich individuals, who were financing stage productions (Vasilescu n.d., pp.1–2). It has no religious background and does not mean in the business world that the angel is not expecting something in return. In the US, the term business angel occurred first in L.A. in the 1920s, when rich individuals helped to produce Hollywood movies (Romans 2013, p.17). BAs contribute a substantial amount to the financing of the startups. Private investors, who invest in unquoted young entrepreneurial companies, are called business angels, because they have an essential role in bridging the financial gap between the financing of family & friends and formal venture capital (Vasilescu n.d.). In addition, BAs offer their experience and network to the startup, which is often even more worth than the money they generate. They bring in their experience from a particular area of business, as well as from their management experience. A study of Olofsson (2001) found out that from all stakeholders it was the business angels, who had been valued by the entrepreneurs, to be the biggest supporters (Olufsson 2001, p.165).

Since business angels are known, due to the individuals joining together, not to be a homogenous group of people, Diane Mulcahy (2005, p.95) therefore classifies business angels in following six categories:

1) The "technology guru": Business angels with industry expertise, attracted by technology and who want to work with the startup to succeed;
2) The "teacher" who is an entrepreneur and behaves like a mentor;
3) The "status seeker";
4) The "investor" who is acting strictly rational assessing the financial numbers;
5) The "portfolio manager" who is a wealthy person – not an entrepreneur, and believes that acting as an angel investor is a smart financial endeavor;
6) The "virgin angel" who has never made an investment and therefore is unpredictable

3.11.1.1 Facts & Figures about BAs

In general, they are investing under $500k in new US ventures and 25k – 250k € in European startups (Saublens & Secretariat 2008, p.10), whereas VCs start to invest above these sums (Osnabruegge 2000, p.92). BAs are sometimes, due to the many similarities with institutional VCs, called "informal venture capitalists" (Macht 2006, p.2)). The market for angel investors is extensive, especially in the US, where angel investors invested about $20bn in new ventures in 2010, in comparison to $17.6bn in 2009 (The Economist 2011, pp.73–75). There are annually very big fluctuations in the VC and BA markets. This is the reason why the 2007 statistics of the BA market might seem relatively large compared to the 2009 statistics of the VC market (see next subchapter). The Centre for Strategy and Evaluation Services (2012, p.5) made in its 2012 report on Business Angels markets and policies in the EU some interesting observations on BAs according to the European Business Angel Network (EBAN). In their report, the average business angel is in 95-97% cases male and between 35 and 65 years of age. They have been previously successful entrepreneurs or businesspersons, who invest between 25k and 250k €. They investment usually no more than 15% of their own funds in a single venture and support the team with their management skills and experience, as well as bring in their own network.

2007	D	EU	US
Estimated Nr. of business angels	5-10k	75k	250k
Average Investment per round in €	100k	165k	210k
Total estimate invested annually in	0,1-0,3bn	3bn	20bn

Table 7: Comparison of the BA Market (Saublens & Secretariat 2008, p.18),
(Centre for Strategy & Evaluation Service 2012, p.15), (European Investment Fund 2011)

The numbers in the above table vary significantly, because the visibility of business angels in Germany is only between 10-25%. Many business angels are not visible and are left out of the calculation, due to the limited knowledge in what they are actually investing. It is not surprising therefore, that different sources show different numbers. Taking a further look into the EU numbers and breaking them down, the average investment per round also differs. This is because one has to distinguish between business angels' average investment round and average investment per business angel per round. In France and the UK for example, the average investment round is much higher than the individual investment per investor. This indicates the building of syndicates, in which many business angels invest together in a startup. Whereas Germany according to the EBAN survey from 2010 (Centre for Strategy & Evaluation Service 2012, p.20) has a proportion of 1,67 investors per deal (191k € deal volume with 114k € per investor), France has a ratio of 9,83 (177k € deal volume with 18k € per investor) and the UK of 5,70 (142.5k € deal volume with 25k € per investor). This pattern for syndicates bears some advantages, because the startup can profit from the experience of several business angels at the same time, while the investors can distribute their risk on several shoulders. If the syndicates are less complicated and easier to build, it makes it easier for more financing of potential startups. Nevertheless, a syndicate can also have some problems when it comes to organizing the deal in the first place and carrying on the venture, when too many investors tell the team what to do. In order to avoid these disturbances, the syndicates usually have a lead investor that helps organizing the deal and the syndicate itself. Often the lead investor also can speak for the others in the shareholder meetings. Business angel networks are also a great help for these syndicates, because it brings many investors together, thus making it easier to form a syndicate.

3.11.1.2 Motivation of Business Angels

Very often BAs have earned and acquired their wealth through entrepreneurial activities or in their past successful business career. They are willing to take these high risks, for several reasons:

1) It is a financial chance and game for them.
2) They receive relatively high equity percentages for their investment, because a startup's value at an early stage is very low.
3) The sums they invest are small compared to their personal assets, thus they can afford to lose it and invest usually in several teams to distribute the risks and chances.
4) It is also an interesting and exciting occupation for them, when they do not want to go into an early retirement.
5) Last but not least, humans like to give their experience to the younger generation; they love to have an audience to tell their business stories and often appreciate the positive feedback they receive from the founders.

The active involvement in a new business, the keeping up with the technological development and the gaining of philanthropic satisfaction by providing advice, as well as making important contacts, is also described by Vasilescu in detail (Vasilescu n.d., p.2). However, Macht (2006, p.3) still asserts the main motivation is the financial investment. It also remains a sort of game for him or her, because the financial outcome is very vague. According to member of the Business Angel Network Deutschland (BAND) many BAs do not receive the financial returns on investment as they have wished for, but still keep on investing and hoping that the next venture will do better. This has some parallels to gambling. Since most groups of BAs do not want to be known and work in public, but rather stay behind the scene with the startup, it is difficult to examine their total success rates and compare them with the original motivation.

3.11.2 Private Investors

Private Investors (PI) are a mix of BAs and VC funds. In contrast to a BA, a PI usually has more than one person investing and organizing the investments. The ticket sizes of the investments often exceed the ones of BAs and range between the investments of BAs and VCs (Manger 2011). Like BAs, PIs are using money from previous business activities, ventures or family assets. Many family offices that manage the private and corporate wealth of a family could be considered part of this group. A major difference to a VC fund, which will be explained in great detail in the next section, is the fact that a PI usually operates with a single source of money it has control over, whereas a VC fund collects money from several outside individuals, corporations and institutions. However, some PIs, that have allocated money for the investment in startups, consider themselves a VC fund and thus overlap with classic venture capital, as it is used in the market and explained in the following section 3.11.4.

3.11.3 Crowd Funding

Crowd funding, crowd lending or crowd investing are particular new forms of funding, in which individuals group their investments together on an online platform, in order to fund startups (Crowdfund Insider 2014). Its general purpose can be investment oriented, sponsorship or a donation. The general term used for all different forms, however, is crowd funding. The beginnings of crowd funding can be traced back to Joseph Pulitzer asking the readers of his newspaper, the New York World, in 1883 to donate money for the Statue of Liberty in New York (Romans 2013, p.28). Nevertheless, the focus of crowd funding on startups and new ideas first appeared in the beginning of the 21st Century based on Internet technology. It then, in 2012, reached its breakthrough with the signing of the JOB act during the Obama administration. Since then, the amount of crowd funding has doubled to over 5bn USD in 2013 with the US accounting for up to 72% of it (Forbes 2013b). On online platforms, like kickstarter.com or seedmatch.de, startups can present their company and claim a certain minimum amount for a specific evaluation, which the startup has set up. Within a certain time-frame, private investors can then buy stakes, through the platform, which organizes the paperwork and often pools the stakes together. The actual legal outcome of the investment could be real equity, preferred stocks, convertibles or other forms of loans. On a donation or sponsorship model, the investors or donors are only getting gratitude or other non-equity/ loan related benefits, such as being among the first group to receive the yet to be developed product. Because of the growing volume, the fact that the startup sets up the valuation and receives publicity through its public search for capital on the platform, crowd funding is an interesting alternative of funding for today's startups.

3.11.4 Venture Capital (VC)

The National Venture Capital Association's (NVCA) definition for venture capital (VC) is *"money provided by professionals who invest alongside management in young, rapidly growing companies that have the potential to develop into significant economic contributions*" (NVCA based on (Orthiese 2007, p.10)). The European Venture Capitalist Association defines it as "*Professional equity co-invested with the entrepreneur to fund an early-stage (seed and Startup) or expansion venture*" (EVCA 2010). There are differences in the usage of the terminology, in- or excluding the expansion financing.

3.11.4.1 Organizational Entity

The typical VC firm is organized as a limited partnership, having two different kind of partners (BVCA 2002, p.4ff). The first types of partners are the limited partners that provide the money to the fund. They usually are not part of the second group that runs and manages the fund. This second group consists of the general partners that have collected the money for the fund and operate it. However, legally they have to operate through a management company that allows them to collect a management fee for running the fund. Typically, this annual management fee lies between 1.5-3% of the total capital fund. In addition, the general partners will receive 15-30% of the profits that the fund generates though its investments and their exits (Gifford 1997, pp.459–460), which is called *carry*. The rest of the profits go to the limited partners.

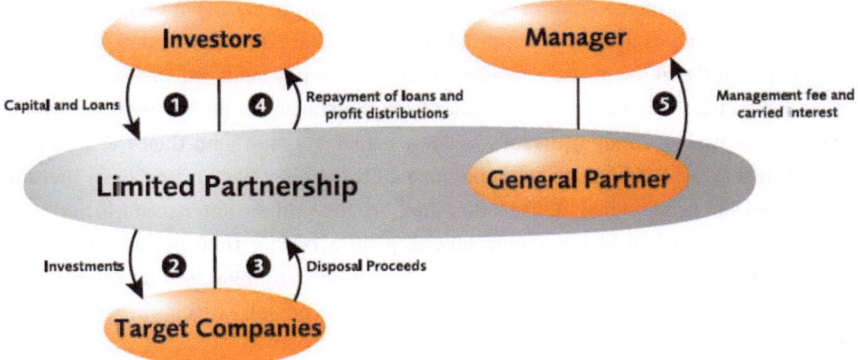

Figure 8: Structure of a VC Fund (BVCA 2012, p.4)

The term *carry* or *carried interest* is said to be derived from medieval merchants in the region between Genoa and Venice that received a 20% carried interest for the cargo they carried, which belonged to others (Romans 2013, p.62).

It should be noted that in a limited partnership, there must be at least one general partner, who is individually or collectively liable for all expenses or legal suits, while the limited partner, under Uniform Limited Partnership Law, is only liable to the extent of his or her investment (Schwarzkopf 2009). That is why technically, many general partners in the VC funds are not private persons but companies like a UK/US limited or a German GmbH with limited liability.

3.11.4.2 Structure and Investment Focus

The money of the VC funds is mainly being raised from pension funds, insurance companies, banks, individuals and other financial institutions (Landström 2007, pp.5–8).

VC companies not only differ in the dimensions of their fund size, their industry and their geographic scope. It is also very interesting for startups to know, at what stages the VC companies start to invest and with what kind of ticket size – the amount they invest per investing round. The age of the fund is also important, because one can derive from it, how many more startups can be financed till the fund's resources end or are limited to already invested ventures. A fund usually runs ten years, and many VCs invest in its first five years in new companies, while in the last five years they invest their money to further financing rounds of the existing companies of their portfolio from the first previous years (Manger 2011; Überla 2011; Romans 2013, pp.62–64). If a fund comes close to the end of this first cycle of five years, it might become very difficult for startups to become the last company they invest in. After the first years VCs begin to collect money for a next fund generation, thus not having the problems of running out of money for new investments after those approximately five years. In a different scenario there could be a fund that has already invested most of its money but is still inclined to take on new companies. This could be problematic for the startup in future financing rounds, when it needs more money, usually even in higher volumes, and then the VC cannot invest in that next financing round, because all the money has already been invested. Very large and professional VC firms therefore, try to avoid the latter scenario and go for the pattern described above – only investing till a certain time period and parallel raising money for another fund before the first fund is running into its second stage, thus having enough money for ongoing financing rounds of existing investments and having money for new startups.

Some VCs focus on particular industry sectors, due to the complexity in different fields of technology and markets. They bring in initial industry knowhow and continue to increase their experience and network through continuous investments in similar areas and startups. The industry knowhow, most VC firms have, is coming from previous experience their entrepreneurs and managers have, and it is also this experience that has helped when collecting the money for the fund. In practice, one speaks about smart money, when investors also bring in expertise and a personal network into an investment.

There is also often a geographic scope, meaning that VCs only invest in companies in a certain area. This could mean a smaller geographic area like Silicon Valley or London Area, as well as an entire nation or international regions (Tyebjee & Bruno 1984, p.1057). This is an important issue for startups, because although many big VC firms have international offices, they look for regional proximity and especially in Europe, they are often looking for national rather than general European proximity. Thus, a startup from Berlin might not get a funding from a London based US investor, who would invests on the other hand in a startup in Chicago, although having an office only in the East and West Coast of the US, with a larger distance than from Berlin to London. Besides the proximity, legal-, language- and market differences are also a reason for a focus on certain regions (Arie 2014).

3.11.4.3 Facts & Figures about VC

When looking at the volumes and sizes of the deals in the US and Germany, it becomes clear that the US exceeds the German VC market by far in total as well as in relation to its larger economy: (US ø size and investment sum p.a. in USD, for Germany in EUR)

	2008		2009		2010		2011		2012	
	US	Ger.	US	Ger.	US	Ger.	US	Ger.	US	Ger.
# Deals invested	2,817	n.a.	2,489	916	3,294	948	3,051	884	3,267	780
Ø Size ($ // €)	11.0m	n.a.	8.59m	0.41m	6.67m	1.0m	10.03m	0.81m	8.63m	0.73m
Invest p.a. ($ // €)	31bn	0,46bn	21.4bn	0.38bn	21.9bn	0.96bn	30.6bn	0.63bn	28.2bn	0.89bn

Table 8: Comparison of the US and German VC Markets (Dow Jones VentureSource 2010) (Jacobi 2013) (Majunke Consulting 2013) (H. Brandis & Whitmire 2011, p.4) (CB Insights 2012) (Bundesverband Deutscher Kapitalbeteiligungsgesellschaften (BVK) 2013) (Bundesverband Deutscher Kapitalbeteiligungsgesellschaften (BVK) 2011a)

Compared to each GDP, the US invests 5,8 time more in VC with a 0,15% of its GDP going into the VC market than Germany with its 0,026% of its GDP going into the VC market (Hendrik Brandis & Whitmire 2011). Due to the much larger economy and some other factors discussed later in this thesis, the total VC numbers in the US are far greater than Germany's. The table above also shows however, that numbers especially in the younger German VC industry differ significantly due to different information, sources and insufficient or incorrect data. Majunke Consulting for example counted 267 venture deals in 2012, with an average investment sum per investor of 3.3m € and a total of 0.89bn €. According to another source, in 2009 the average yearly investment per investor was only 0.4m € leading to a huge spread between 2009 and 2012 of almost 3m

€ (0.4m to 3.3m). The increase of the total market from 0.38bn € to 0.89bn € can in parts be explained by the recession during the financial crisis prior to 2012. The amount of deals however seams to be very low in comparison to 3.5 times more venture deals in 2009 counted for by Jacobi (Jacobi 2013). One would expect the opposite with fewer deals in 2009 in comparison to 2012. The BVK (Bundesverband Deutscher Kapitalbeteiligungsgesellschaften) mentions 780 deals for 2012, which appear to be more accurate and therefore, have been used. But even these huge differences in the German data do not change the fact that the German is way behind the US venture market, having less available information, accurate numbers and statistics.

3.11.4.4 Differences between BAs & VCs

Since the VC stands also for the group of highly risk investments, one needs to discuss the differences between business angels and venture capital funds, which are part of the general asset class of highly risk investments.

Osnabruegge (2000) has analyzed these differences between business angels and thus refers to them as "Institutional Venture Capitalists". He uses an agency theory to point out that the main differences between the investment cycles of a VC and a BA appear from their monitoring and control of their portfolio companies. Both try to reduce the agency risks during all investment stages, but a BA focuses more on the *ex post* investment (influence and work after the deal), whereas a VC does it *ex ante* the investment (influence and work before the deal).

The BA basically invests his own money and usually is more involved in the operations than a VC. Thus, the BA he can have more influence *ex post*, due to his involvement in the daily business after the investment has been done. A classic principal-agent relationship between the entrepreneur and the investor occurs. It should be noted that the VC acts as both, principal and agent at the same time: principal of the portfolio company and an agent of the investors of the fund (the limited partners) the fund's investors. The goals of the limited and general partners might not be always aligned, because of the *carry* and the management fee that had been described earlier. It can incline the fund manager sometimes to invest in high-risk assets instead of safer ones, in order to receive a higher *carry* without bearing the risk of a total failure.

Two cases of an agency problem can be seen according to Osnabrugge (2000, pp.93–94):

1) Conflicts in alignment and verification of goals and

2) Conflicts of risk sharing.

Unlike the BA, the VC must show a clear and stringent plan already from the beginning of the investment process, when collecting the funds. He or she supposed to have a good reputation and promising investing scheme in order to convince the potential investors for the fund in the first place. He or she needs to invest according to the pattern proposed to the investors of the fund. That is why the VC acts *ex ante*, because the major influence on the startup – the actual funding and reputation received through the VC's own reputation – as well as the scrutiny of the startup and decision making on the funding is done before the deal. Modern VC, however, also tries to offer more value than money and reputation, by bringing in their network or guiding the startup through advice and a board seat. This *ex post* involvement is often connected with the term "smart capital", standing for an investor, who provides more than just capital.

3.11.5 Series A, B, C

The broad range of a VC funding and the continuous development of a startup also leads to a certain cluster. Especially in the US, it is common to categorize the investment size and time after the first investment - the seed investment - into so-called Series A, B, C etc. (Investopedia 2014; Newton 2001), in which only external investors are involved. The seed phase can, however, involve family and friends as well as many business angels, and thus also involve external investment. Before the Series A, a startup is supposed to earn first revenues and to show some "proof of concept" however, it may be still far away from profits and success. The more the business model is proving its potential, the more and fast development of the company may require extra money. This stage will then lead to a Series B funding. Depending on the business and industry the startup is in, a further Series C or even D, E, F etc. is needed to expand and grow at a high velocity. The startup could already make profits, but to succeed for example in foreign markets, further funding is necessary. Or in cases of Internet marketplaces, like Amazon, it may require venture funding over many years – besides high revenues in the 3 digit millions – due to the lack of profits arising from necessary and constant investment into the market development and penetration. The average size of the Seed investing and Series A,B,C etc. is hard to define, yet a few sources reckon Seed rounds in the US to be less than 1.5m USD (CB Insights 2012).

The author has derived the following table, based on an analysis of the venture capital deal database of crunchbase.com (Crunchbase 2014). Since many business angel deals are not officially known, many deals do not show up in the Crunchbase database. They might be part of a Series A in the lower segment of less than 1m USD or in the Seed phase. The following table therefore points out the financing areas of Series A and B. The sum of the investments from the Crunchbase database for a Series A and B in the year 2012 for example were over 14bn USD. According to Preqin, the global VC volume in 2012 with 5,005 deals was 39bn USD, with Series A and B deals accounting for approximately 20% of the deals (Morris 2012)(Preqin 2013, p.3). However, the Preqin data collection lists 35% of the 5,005 deals to be not specified rounds, leaving 1,750 deals open to belong to various founding rounds. Therefore, the Crunchbase dataset can be used as a statistically valuable source, with its more than 1,607 Series A and B deals.

Series A (in USD) in	2013	2012	Series B (in USD) in	2013	2012
< 1m	100	86	< 1m	17	0
1-5m	667	553	1-5m	210	161
5-10m	318	222	5-10m	210	154
10-20m	131	146	10-20m	267	158
20-50m	49	50	20-50m	92	66
50-100m	3	3	50-100m	3	8
>100m	0	0	>100m	6	0
Sum of deals	1,268	1,060	Sum of deals	799	547
Estimated Average	6.32m	6.84m	Estimated Average	12,16m	12,77m
Top Areas (1-5m)	53%	52%	Top Areas (10-20m)	33%	29%
Top Areas (5-10m)	25%	21%	Top Areas (1-5)	26%	29%

Table 9: International Series A & B Deals 2013 & 2014 (Source: Author & Crunchbase 2014)

The data indicate that more than 75% of Series A investments are in the range of 1-10m USD and 50% in the area between 1-5m USD. Taking the medium value of each area (e.g. 2.5m for the 1-5m area) for the years 2012 and 2013, it will lead to an average volume of a Series A at around 6.5m USD (compare data and intelligence from Preqin (Morris 2012) counted 6.7m USD in 2012). The Series B are about double the size with an average at around 12.5m USD (compare data and intelligence from Preqin (Morris 2012) counted 11.8m USD in 2012) with the 10-20m range accounting for most of the deals over the two years. In 2013 Crunchbase listed 29 deals between 50-100m USD in its Series C section, showing the further increase of ticket sizes in each Series. The total amount of deals on the other hand decreases to only 328 in the range from 1-300m USD for a Series C.

3.11.6 Comparison of Different Risk Capital Types

An overview and summary of the different setups and characteristics of risk capital are shown in the following table. The mentioned stages of a startup can here also be related to the Seed financing and Series A,B,C categorization. The Early Stage can be compared with the Seed financing phase and an early Series A type, whereas in general Series A and B are more related to the growth phase. In general, the VC players are not strictly bound to certain phases, ticket sizes or rules, and there is no official agreement on these phases nor on the descriptions of the players. Some business angels consider themselves a VC, some private VC investors reckon themselves to private equity investors, while other VC investors say they invest in Early-Stage startups. Interestingly enough, the German investing scheme eventually shows that the German investors are actually investing much later than the Americans do. Once again, there are differences between the US and Germany, with the US having already many VC funds that start with investments in startups that have not entered the market and therefore, are called pre-revenue and sometimes even pre-product investments (Business Insider 2011). Many of these fall in the Seed category, but these Seed or even Pre-Seed financing rounds are often not listed, thus making it very difficult to officially compare the German and US numbers in detail. Informal interviews with several VCs and BAs in Germany and the US have also supported the assumption that US investors start earlier investing in startups than German investors do.

Type	Common Ticket size[1]	Stage	Origin	Motivation
Business Angel	- 10k-150k	- Seed to Early Stage - Seed phase, early Series A	- Private - Corporate	- Personal motivation - Alternative investment
Private Investor/ Fund	- 0,1-0,5m[2]	- Early Stage to Growth - Series A	- Private - Family offices	- Alternative investment - Diversification - Personal motivation
Venture Fund	- 0,5m-20m[3]	- Early Stage to growth - All Series	- Pension Funds - Private - Family office - Corporate - Additional public means	- Investment instrument - Profitable alternative - Strict time schedule & ROI target
Public Venture Fund/ instruments	- 0,1-5m	- Seed to Growth - Series A, (B)	- Public money - Additional corporate means	- Fostering the ecosystem - Creating jobs - Alternative investment
Private Equity deals	- 5-550m	- Later Stage - Higher Series	- Private - Corporate	- Alternative investment - Diversification - Private wealth management
Crowd Funding	- 50-150k aggregated (0.5-10k for each individual)	- Seed to Early Stage	- Private	- Participating in startup business with small personal ticket sizes

Table 10: Comparison of Investors (Source: Author & (Crunchbase 2014))

---

[1] Volumes in the US often double the size in comparison to Germany. The table indicates the volume sizes often seen in the market not an average or exclusive range. Numbers are a mix of EUR and USD

[2] Very few funds in this category, which is often considered the valley of death, because startups seeking money in that range will have great difficulties. The size may differ dramatically, if a private or corporate investor invests heavily in one concept or startup e.g. Michael Otto was said to invest more than 100m € in a new E-Commerce startup for the Otto Group called Mary & Paul.

[3] Especially in the US there are also funds and many rounds that can carry out financing rounds between $10-100m and more, which often enter the stage of private equity also explained in this thesis. Especially for industries like pharmaceuticals, biotech or real estate the investment sums could be much higher.

The amount of different risk capital with different investment schemes is also crucial for the entrepreneur and is an essential part of the ecosystem. The founders need to understand the motivation and potential of each investor, in order to find the theoretically best match for them. In practice, an underfinanced market like Germany, however, will lead to a different reality, in which the founders will not have the luxury of such a choice. This will later be explained in chapter 5.2.1 of the buyer's and seller's market. Most VC funds have a higher time pressure to exit out of the startup due to the promise to their fund investors to double their money within 10 years (Überla 2011; Romans 2013, pp.62–65). A private investor, who mainly operates with money from a single origin, is not under such time pressure. Businesses that require longer periods for development or teams that would like to build up a company without any exit plans to sell parts or their whole company, will have to make greater efforts, in order to convince VC funds to invest.

3.11.7 The Right Fund Size

The ticket size also correlates with the fund size. Due to the risk of underperforming or total loss of startup equity, funds invest in many companies, in order to have a diversification of risk.

3.11.7.1 Fund Size Calculation

A fund that wants to start with a 1m USD for each first investment might than need 10m USD for 10 startups to start with. If it decides to further invest in only half of the startups, in a Series B with 2m USD, it will need another 10m USD. This already adds up to 20m USD, but since the investment period for those Series A and B startups might take only 2-4 years, the question is whether to invest in more startups for the next years or whether to have further money for continuing rounds and/ or higher "ticket sizes".

In addition, the fund also has to pay the investment managers fixed salaries in form of a management fee between 1.5-3% (Romans 2013, p.62; Timmons & Spinelli 2004, p.243). In the case of a 20m USD fund, this would mean at least 200k USD annual management fees. In ten years it would cost the fund a total of 2m USD just for the management fee. But since the screening, investment and management process for the 10 invested startups, from the example above, takes time, it is unlikely that only one person is in the operation involved. Large funds in Germany are said to invest only in 1 to 3% of the business concepts they get acquainted with. In the used example it would mean a screening of 300-900 concepts. The funds, therefore, rather have 3 to 4 investment managers plus assistance, leading to annual management and overhead costs of up to 1m USD per year. This team size, however, makes only sense with more investments over the 10-year period. Therefore, a rule of thump for a strong VC fund is to have a volume of 100m USD or more (also see Brandis & Whitmire 2011 p.35). A management fee of just 1.5% would still leave 85m for investments in a ten-year cycle. In the case of a 100m USD for example, the 85m USD could be distributed on 25 startups with 1m; 12 with an extra 2m and 6 with an extra 6m USD then. 4 investment managers would then handle about 6 startup cases intensively, plus the screening, negotiations and some organizational aspects within the fund.

Having examined the ticket sizes that are being paid (Table 9: International Series A & B Deals 2013 & 2014 (Source: Author & Crunchbase 2014) in a Series A and B, even a 100m minimum fund size appears to be small. Most cases, from the Crunchbase database, are from the US, which has not only 227 funds with a volume of more than 100m USD (Hendrik Brandis & Whitmire 2011) but many of these also have a fund size over 100m USD – up to a billion USD and more e.g. Andreessen Horowitz, Sequoia, Accel Partners, Kleiner Perkins etc. (Wallstreet Journal 2014). In fact, Spinelli and Adams (2012, p.403) list the structure of VC in the US by volume done by the National Venture Capital Association, showing that in 2005 there were even 106 "Megafunds" with a fund size higher than 500m USD and 76 "Mainstream" with a volume between 250m-499m USD.

3.11.7.2 Importance of the Fund Size with the Example of WhatsApp

A recent success case, in 2014 in terms of financing and exit, is WhatsApp, which has been acquired by Facebook in February 2014 for 19bn USD (Forbes 2014b). It is a good example for comparing US and German financing. The seed financing in 2009 of 250k USD (Crunchbase 2014) could have been paid by German investors, but probably would not have been executed, because of the business model that required a huge global network effort, and because it did not bring in money in the first place, due to free usage in the first year and a super low price for the app at around 1 USD for following years. The next and first real serious round of 8m USD would have caused a problem for all German funds, due to risking so much money in proportion to the fund size at such an early stage. The following and final round of 40m could not have been financed by any German fund alone. In contrast, Sequoia Capital from the US funded WhatsApp by itself. They are one of the US oldest and renowned VC companies and have for example raised in 2013 1.17bn USD for a new fund (Reuters 2013b). This potential is important to understand for analyzing and improving the Entrepreneurial Ecosystem.

3.11.8 Brief Outlook on the Development of the Venture Capital Market

Modern venture capital is said to be born after World War II, when the Harvard professor and French general Georges Doriot incorporated *the American Research and Development Corporation* (ARDC) together with Karl Compton, president of the MIT, Merrill Griswold, chairman of Massachusetts Investors Trusts, and Ralph Flanders, president of the Federal Reserve Bank of Boston. ARDC did fundraising from rich individuals and college endowments, and then they invested the money in young entrepreneurial startups, mostly in the high-tech manufacturing (Botazzi & Da Rin 2002, pp.233–234).

The VC funds spread from the Boston area to California during the late 1950s. Funds were established in San Francisco and one of its suburbs - the Silicon Valley. The VC industry grew steadily, until the beginning of the 1980s. At that time, some new tax incentives gave the pensions funds the ability to invest in these funds. As a result, money started flowing rapidly and in large volumes into the funds.

VC was more or less an American institution, until the late 1980s, when the European VC started to develop rapidly. This was made possible by an introduction of secondary stock markets in many countries, which allowed smaller firms to make IPOs, and thus generate a further interesting exit option for investors (Landström 2007, pp.13–14). In the 1990s the Asian VC market tipped-off, with many American VC companies opening offices abroad. Today, the VC industry has spread throughout the modern world, starting also in the BRIC (Brazil, Russia, India and China) and emerging markets (Botazzi & Da Rin 2002, p.234); (Landström 2007, pp.12–15). The sums invested in China for example increased from $1.8bn to nearly $5.4bn between 2006 and 2010 (The Economist 2011, pp.73–75).

3.12 Investment Decisions of Venture Capitalist & Business Angels

VCs and BAs play a significant role in providing capital to startups. In order to understand the mechanism of risk capital financing, it is important to understand their decision making process. A VC uses various criteria when making an investment decision. The main goal in this decision making process is to get sufficient information in order to reduce the risk of adverse selection and to *"separate the wheat from the chaff"* (Fried & Hisrich 1994, p.29).

3.12.1 Adverse Selection at the Beginning of the Encounter

A general information exchange takes place in the beginning. Adverse selection, which is sometimes referred as negative selection, occurs in this context from an information asymmetry with a set of problems between two parties, the principal and the agent. In those cases, where the investor acts as the principal and the entrepreneur as his agent, the investor must keep in mind that the entrepreneur has some exclusive information about certain matters, which affects the valuation of his or her investment. George Akerlof (1970) in his article "The market for lemons" called attention to the importance of the agent's information, using insurance contracts as an example. Akerlof's observation can also hold true and be applied in the startup world. If the entrepreneur has crucial information about potential problems that could endanger the company, then the principal investor, who also carries a risk in financing the company, must be informed of the crucial information. This is an agency duty, which applies to investors, but not to banks, which are not principals, because they only lend money. Companies will ask Banks for classical loans, when they can to a certain extent, assure success and low risks. Startups on the other hand, one can assume, must look for BAs and VCs – Akerlof's lemons – when high risk is involved. Thus, it is quite possible that these type of investors always end up with the more risky and uncertain companies, whereas banks only get to "good" ones. In practice, this generally does not occur, because on the one hand, risk capital is most of the time a limited factor, in which experienced managers or entrepreneurs screen the startups, compare them and select the most promising ones. On the other hand BA and VC businesses are based upon risk, which is compensated with high equity percentages. In addition, good VCs believe to know the markets better than the startups and can even influence them positively or improve the success cases through their network and contacts. BAs on the other hand become sometimes part of the operational team and thus can improve the startup's chances. Especially VCs protect their money with complicated contracts and profound due diligence when necessary, thus it is hard for startups to hide crucial parts.

3.12.2 Investor's Decision Process

The VC's decision process has been described in several works of Wells (1974), Tyebjee & Bruno (1984), Fried & Hisrich (1994) or Boocock & Woods (1997). One of the first models investigating the VC investment decision process and its linkage to the venture's performance was the model described by Tyebjee & Bruno (1984) in their often cited paper "A Model of Venture Capitalist Investment Activity". Of course, there had been already some studies before about the process itself. For example Wells (1974) conducted a study about the screening process of VCs, but what makes the study of Tyebjee & Bruno special, was the combining of the screening criteria research with the perceived risk and expected return of the ventures and so indirectly with the expected performance of the venture. They identified four phases:

1) Deal origination
2) Screening

3) Evaluation
4) Deal structuring.

The study by Fried & Hisrich (1994) concentrated also on the investment and evaluation process. They divided the screening phase into a firm-specific and generic screening, and the evaluation into 1st and 2nd phase evaluation, thus ending up with having six different phases.

Boocock & Woods (1997) had a slightly different approach. Their decision process consisted of 7 phases, and especially added two meetings and presentation by the founding team to the core phases of the evaluation, and thus required more detailed on the screening:

1) Generation of a deal flow   5) Board presentation
2) Initial screening   6) Due diligence
3) First meeting   7) Deal structuring.
4) Second meeting

Throughout the stages, VCs try to consider the success and failure chances. They use according to Payne (2009, p.157) different criteria such as

- Skills and capabilities of the entrepreneur/team
- Product and service characteristics
- Market characteristics
- Potential financial returns

Nonetheless, the basic structure of the process seems to be similar in each study. They all agree that there are multiple stages in the process and that at least two stages 1) screening and 2) evaluation (Hall & Hofer 1993, p.29) must exist.

Step	Tyebjee & Bruno	Hall	Fried & Hisrich	Boocock & Woods
1.	Deal Origination	Generating a deal flow	Deal Origination	Generating a deal flow
2.	Screening	Proposal screening	Firm-specific	Initial screening
3.	n/a	Proposal assessment	Generic screen	First meeting
4.	Evaluation	Project evaluation	First phase evaluation	Second meeting
5.	n/a	n/a	n/a	Board presentation
6.	n/a	Due diligence	Second phase evaluation	Due diligence
7.	Deal Structuring	Deal structuring	Closing	Deal Structuring

8.	Post Investment activities	Venture operations	n/a	On-going monitoring of investments
9.	n/a	Cashing out	n/a	Cashing out

Table 11: VC/BA Cycle (Boocock & Woods 1997, p.41)

3.12.3 Venture Capitalist & Business Angel Cycle

There are many similarities between the investment procedures of a VC and a BA, but there are also some significant differences. Both cycles expect a high return on investment during the period of five to ten years. Rule of thump for VCs says, the minimum is doubling the fund's money within 10 years, bringing about at least 6-8% increase p.a. (Romans 2013, p.62). BAs & VCs normally invest only in areas, where they have industry know-how. Whereas the BA needs much more geographical proximity due to his or her operational role, the VC places more emphasis on the availability of services close by, where the portfolio company exists (Ehrlich & De Noble Tracy 1994, pp.69–70). Ehrlich & De Noble Tracy also mention that the invested amounts differ by tenfold or more; VC's are investing $1-3m (in the first round) in contrast to the $0.1-$0.5m of the BAs.

3.12.3.1 The Deal Origination

The deal origination phase is the first in the investor's decision process. It is actually the phase, when the investor gets the initial information on the investment target. Awareness of potential deals can be divided in three groups. First group are the so-called *cold contacts*. In that case the entrepreneur contacts the VC, who does not know the entrepreneur. The second group is about startups and deals the VC and its staff scout and find for themselves, through a smart network, which they usually have built up, and through their access to promising teams and sources. This network also brings in the third and often most promising and preferred group (Perkins 2007) – recommended by trusted people out of their network. While *cold calls* seldom end up as a deal, referred proposals on the other hand make up 60-80% of the closed deals (Larsson & Rooswall 2001, p.30). Fried & Hisrich (1994, p.32) listed common referral sources, which included investors in the VCs' fund, investment and commercial bankers, management of the firms in the VCs' portfolio, consultants with a working history for the VC and family and friends.

3.12.3.2 The Screening

After being offered deals, the VCs start the screening process. This phase can be divided to two parts; VC-specific screening and the generic screening. Screening means in this context that they sort out the propositions, which do not fulfill their screening criteria.

According to some experts, the screening phase is not a democratic or balancing process, for instance an unacceptable value in one part cannot be compensated by a high value on other part (Riquelme & Rickards 1992, p.505). The venture team cannot afford sending a poor business plan (BP) to the reviewer, because the number of BP received by a standard VC firm in Germany exceeds more than 2,000 a year (Nagel 2011). Of course, this includes also a lot of *cold calls* that do not receive a high attention. The firm-specific screening phase is more about the suitability for the specific VC firm. Criteria, like general market size, company stage, experience or location of the business and the sector of industry will be reviewed. Therefore, most deals are rejected without analysis in this phase (Hall & Hofer 1993, p.28). A study of Hall and Hofer found out that VCs initial screening took about 6 minutes, and in the initial screening the key criteria included the fund's lending guidelines, the long-term growth, as well as ROI of the industry sector. VC fund managers usually need to follow their decisions, because they have sold their criteria to the fund's investors. On the other hand, also the entrepreneur's experience and the existence of a product, as well as the product features have shown to be part of the decision making (Riquelme & Rickards 1992, p.506). The next step in the process will be more a generic screening.

The generic screening concentrates on the business plan, again rather than finding reasons for a rejection. The majority of the deals will also be eliminated in this process. One goal is always to eliminate the least promising deals, which also includes sometimes loosing a few good ones.

3.12.3.3 The Evaluation

The VC will then start to gather information about the filtered ventures. The first evaluation is the start with the due-diligence, which is a classic instrument in the banking scene used for mergers & acquisitions (Gabler n.d.). It begins with a meeting with the management team of the startup. The VC not only wants to learn more about their business, but also wants to get to know the founders and core team better, in order to evaluate their quality and abilities (Fried & Hisrich 1994, pp.31–32). VCs need to get a positive impression of the entrepreneur, and what matters the most for everyone is the competent and team (Author's experience in meetings with very successful entrepreneurs and investors. Also compare (Nagel 2011)). Besides several meetings with the team, the evaluation includes visits to the facilities, speaking with referrals and former colleagues of the team, exchange information with other investors, interviews with current and potential customers, as well as an analysis of the market (Fried & Hisrich 1994, p.33). Besides the team evaluation, the approach in this phase is very analytic and tries to cover the KPIs and important product, as well as market data.

If a market already exists and it is possible to get enough information on it, market studies must be made. The sustainability of the competitive advantage also plays a very important role. Furthermore, part of the analysis is the threat of market entries by new players, the development of substitute products and the IP protection of the venture (Silva 2004, pp.138–140). If the market does not exist, potential customers are a source of information, in order to get a feeling for a market acceptance. With Early-Stage investments, technical studies are more important than with Later-Stage investments, because they have already customers and industry experts one can discuss the technical features with. With Early-Stage investments, a technical evaluation is usually done with the help of consultants or technical advisors of the fund managers (Fried & Hisrich 1994, p.34). The existence or potential for an IP protection does play an important role when reviewing the product or startup in this phase (Riquelme & Rickards 1992, pp.505–506), because it stands for sophisticated and new technology on the one hand, and for a market barrier for competitors to enter the market on the other hand.

VCs and especially BAs like to talk with their peers about potential investments and often form syndicates, in order to invest as a group, share experiences and risks. The projected value of the company is also calculated or estimated with the given information and compared to other companies of the industry sector. For startups these estimations are very difficult to understand and driven by the investors experience and often also gut feeling. For later-stage investments with a financial track record, the evaluation becomes more transparent and meaningful, because a financial stream can be discussed and developed (Fried & Hisrich 1994, p.34).

Ideally after the first evaluation the investor sees already the potential of the company and has hopefully developed an *emotional* commitment to the proposal (Fried & Hisrich 1994, p.34). This means that if he goes into a second phase, the VC has a serious interest in the startup and will from there on try to close the deal, as soon as possible and eliminate remaining obstacles. The line and time between the first and second phase of an evaluation can be very thin and short, thus Tyebjee & Bruno (Tyebjee & Bruno 1984) did not even have a two-stage evaluation phase in their model.

3.12.3.4 The Deal Structuring

When the decision to invest has been made, it can still fail if the entrepreneur and the investor do not agree on the deal structure. The crucial issue is always how many and what type of shares of the company the investor will receive for his investment (Tyebjee & Bruno 1984, p.1051). Other issues are about the protective points the investor wants set up in order to protect his investment. These can include: Vesting periods for the founders, good and bad leaver agreements, limiting capital expenditure or fixed and variable management salaries. They also set the basis for the mechanisms, how the venture capitalist can take control of the board, or force or profit from liquidations (Drag- and tag along agreements, liquidation preference) even if the VC is holding a minor position in the company (Tyebjee & Bruno 1984, p.1054). A performance-dependent mechanism, which makes the entrepreneur's share on the company dependent on the performance of the company, is also a common instrument of control. This can even lead to a change of the management team, if the investor has a majority. Despite the tremendous effort in evaluating the company, according to Fried & Hisrich (1994, p.34) only about 20% of the companies, which have made to this stage, will finally receive financing, after the successfully completed deal structuring phase.

3.12.3.5 Post-Investment Activities

After the deal has been made, the supporting and monitoring functions of the investor will be set in place. There is a principal difference in the post-investment involvement of a VC or BA and a classic financial intermediary such as credit institute or a "normal" stockholder. Whereas the banks and stockholders normally have a passive role, which includes mostly the monitoring of the firm, VCs & BAs actively participate in the decision-making and parts of the operations. Monitoring is done to prevent and to decrease the moral and legal hazard of the agent (Landström 2007, p.193). Value-adding is done to enhance the chances of success of the company and thus improving the chances for high ROIs (Luukkonen & Maunula 2007, pp.3–4).

3.12.3.6 Monitoring

As the goals of entrepreneurs and investors may not be always aligned, the investor has a special motivation to observe and control the actions of the entrepreneur. For instance, the entrepreneur might be more interested in building a great working atmosphere and solid business, while the investor is always looking for a profitable gain or high exit.

Monitoring in the post-investment phase may be conducted in several ways. The classic way of monitoring is taking a seat in the board of directors of the company, having veto rights and being able to make propositions. The investor may also bring an external expert to the board, if he himself does not have the time or lacks a certain expertise (Luukkonen & Maunula 2007, pp.17–20). The formal ways of monitoring are followed by informal ways like monitoring the financial and performance reports of the company, visiting the company occasionally (Landström 2007, p.197) and Jour Fixes via telephone, email or personal meetings. Monitoring increases the agency costs on both sides: the entrepreneur must produce the information and make it available to the investor. The investor must spend time and put effort acquiring and interpreting this information. The amount of time the investor is able to devote to each startup depends on the number of companies in the portfolio, the importance of active monitoring in the individual case and the operative setup of the fund. However, research believes that the VC monitoring leads to positive outcomes both for the portfolio company and the stakeholders (Landström 2007, p.198).

3.12.3.7 Value Adding

Besides making sure, that the entrepreneur will not make the general management mistake or anything harmful for the fund's money, VCs can contribute their know-how to the startups. The first influence might have started already before the deal, when the startup had to develop and adapt its operational planning, its templates and calculations in accordance to the investor's standards and expectations (Luukkonen & Maunula 2006, p.4). However, most of the value-adding can be done after the deal.

Landström (Landström 2007) classified three ways, how investors add value to their portfolio:

1) Strategic help e.g. through involvement in the board.
2) Provision of key external contacts for recruitment and professional services, giving access to key customers and connecting to additional finance sources.
3) Personal consulting, mentoring and coaching (Landström 2007, p.201; Luukkonen & Maunula 2007, p.3; Gorman & Sahlmann 2010, p.237).

Fried & Hisrich (1994) divide VC value-adding activities even in 6 different parts:

1) Operating services: consulting, funding, partnership arranging, advising, guidance, crisis management etc.
2) Networks: Provision of contacts for funding, underwriting, board of directors, management team, service providers, customers, suppliers and corporate partners.
3) Image: the VC involvement signalizes quality and gives the stakeholders confidence on the company. This reputation helps the company to attract talent, money, customers and "air time" to their business.
4) Moral support: discussing partner on sensitive issues or during a crisis.
5) Knowledge about business in general.
6) Discipline: VC puts pressure on the investee to achieve the objectives with stick (threat of management change) and carrot (better contract terms) (Macht 2006, pp.3–4). Additionally, Maunula & Luukkonen (2007, pp.23–26) reports that VCs also give the portfolio companies marketing assistance.

Even though the investment cycles are quite similar between the VCs and BAs, differences in their value adding do exist. The entrepreneur should calculate the benefits and costs of the particular investor to choose a successful deal. Even though the BAs are more heterogenic group than the VCs are, some general differences in their involvement can be found. Macht (2006) has listed some of these differences. First of all, BAs invest in earlier stages than VCs, where the management lacks experience and market acceptance The BA can share his or her experience, find first customers, motivate the team through his early investment and by his or her confidence and eventually help in convincing further investors to join. (Macht 2006, pp.5–6) At that stage, the BA is almost part of the management team and its daily work life. It is not uncommon that BAs commit 1-2 days per week for each of their investments. VCs would normally get only involved in management activities, if major problems occur or key accounts need their presence. Some BAs still will try not get involved operationally at all, because they either have too many investments or rather want to invest and coach teams than work in the business. VCs are sometimes also more eager to replace the management team than the BAs would, thus sometimes causing problems and differences between investor and entrepreneur to arise. There are also differences with the time frame of these investors. VCs normally have formal rules when to exit the position, in a pre-written holding time or closed-end fund, whereas BAs do not. This gives the VCs more time and leverage for an active involvement (Macht 2006, pp.3–5). Eventually, due the stage of investment and the ticket size being paid, BAs and VCs are not alternatives for each other. A VC investment might rather follow a successful BA investment that has helped to increase the value of the startup to make it eligible for a VC investment.

3.12.3.8 Exit

In the exit phase, the venture cycle will come to an end. As especially VCs have a certain ROI opinion, which they have promised to give their fund's investors. If they want to sell their shares of the startup, the parties must do so at time, in order to fulfill the raised expectations.

There are basically five alternatives for the investor to exit out of company (Cumming & MacIntosh 2003):

1) IPO (Initial Public Offering)
2) Acquisition
3) Buyback
4) Secondary sale
5) Write-off

From these alternatives, the IPOs and the acquisitions are normally the most profitable ones, and these are preferred by venture capitalists.

IPO means offering the shares of the company to public investors. The IPO market was very attractive in the late 1990s, when young IT firms were booming, going public and money was easily available. The IPOs slowed down significantly after the burst of the New Economy Bubble (also referred to as dot.com bubble) in 2000 and to this day scars in the EU and German market exist. There was especially a high skepticism for young technology companies going public. Nevertheless, the billion dollar IPOs of Facebook, Twitter or Groupon thereafter, have helped the US market to recover much faster from such a downturn, than other world markets (Hofmann 2014; Beattie n.d.). This is true also for those US companies' offerings after the recovery from the 2008 financial crisis.

A VC will in general not sell all of his shares of the IPO, and neither will the entrepreneurs. A full or partial exit will take up to one year after the IPO, in order not to upset the market and keep the stock price high. VCs sometimes try to buy a premature IPO, in order to build up their reputation and to demonstrate a strong performance of their fund. That becomes necessary if the VC managers are about to collect more money or open up a new fund (Landström 2007).

When a third party purchases the entire company, it is called an *acquisition*, which could be "standalone" or as a consolidation or merger with another company. The buyer can be the mother holding of one of the investors, a customer, a supplier or even a competitor. The basic ideas behind an acquisition are synergies, the gain of key technologies or customers, as well as the elimination of an important competitor (Cumming & MacIntosh 2003). Some financial investors belief, they can add parts to the or sell of parts from the company, and thus increase their total investment value. Even at this stage, an acquisition is not only about the achievements, but also about huge potentials and phantasy the company and the business model have. This is the case in most mergers and acquisitions, no matter how big and mature the companies are (Deans et al. 2002).

When VCs sell their shares to a third party while the entrepreneurial team still keeps their shares, it is called a *secondary sale*. The third party could be another VC fund or another company (Cumming & MacIntosh 2003).

On the other hand, a *buyback* or *management buyout* is the purchase by entrepreneurs or firm managers. VCs often have a paragraph in their contract, which allows them to sell the shares at a certain price after a certain time interval, if the firm has not gone public, was not acquired or did not achieve some other set of goals (Cumming & MacIntosh 2003). To avoid a secondary sale, the entrepreneurs and managers of the companies can try to work out an agreement with the investors on the acquisition of the investor's shares.

The worst scenario for an entrepreneur is the *write-off*, which normally is a total or part failure of the venture (Cumming & MacIntosh 2003, pp.8–9). In this case, VCs might also keep their shares of the company, hoping that the market might turn, or the team will be able to do a turnaround. Thus, a write-off is avoided but a point has been reached, where investors have only few options for financial returns or a later exit (Ruhnka et al. 1992, p.1). At this point, they do not expect to receive revenues from dividends or a sale of their shares.

3.13 Business Planning

A business plan (BP) is a formal and brief description of the setup and future of a business. The plan covers many important "W-questions", like *what* kind of product or *what* service is planned to be launched, *who* are the customers, *why* will they buy something from the company, *how* shall it be done, *when* will it be introduced in *which* market, in *what* timeframe and who is part of the team (U.S. Small Business Administration n.d.).

Depending on the stage of a company, different aspects and levels of detail are required. There are BPs for teams that are about to start their venture; there are templates for SMEs and even large corporations require general forms of business planning and specific ones for their projects. Detailed business plan documents are a standard in the industry and essential for mergers and acquisitions (UBS n.d.; Business Net Partners 2013). For this thesis however, the emphasis lies on the BPs for startups.

BPs are intended to list, to describe and to perform a number of tasks, not only for external parties, but also for the team that writes it (Entrepreneur.com n.d.; Romans 2013). They are mainly used to convince investors, set up deals with potential suppliers, attract key employees with a promising development and for the team itself, to understand how to implement their concept and manage the company better (Romans 2013, p.104).

A BP points out the goals and helps to derive the strategy how to reach them. It also deals with the challenges the business is confronted with and the potential approaches that can be taken. Writing a business plan also demonstrates dedication to investors, as well as stamina and sincerity. It usually takes several weeks to complete a first decent version. The BP can also show structure and intelligence of founders, when it comes to solving problems and answering questions like market size, business modeling or scenario planning. Although there are many standard approaches, each business requires tailor-made suit and a smart adaption. The W-questions will eventually be used to create a business case, which calculates the financial development of the company ideally in several scenarios. With that, it will be possible to estimate the necessary investment and liquidity.

3.13.1 The Basic Structure of a Business Plan

According to many sources, good business plans follow generally accepted guidelines for both, form and content. There are three primary parts to a business plan (Entrepreneur.com n.d.):

1) The business concept is discussing the industry, the business structure, the particular product or service, and how it will become a success.
2) The market section is describing and analyzing potential customers. Who and where are they, what makes them buy products and services, how can they be reached and how can they be kept. It is also about describing the competition and the positioning of the company to be competitive.
3) The financial section is containing the business model, the income and cash flow statement, the balance sheet and other financial KPI[4]. If financial help from outside is needed, an investment case needs to be added.

Struck (1990) has looked from another angle and thus structures a BP in three other areas, which are also accepted in the German VC community, however do not contradict the three primary parts described above:

1) A description of the situation, how everything is connected, what assumptions are to be made and what is planned is the starting point for Struck.
2) Numbers and calculations are bringing the assumptions and plans to a quantitative level. From revenues, over liquidity up to KPIs like number of employees needed.
3) The annex completes the set with important market studies, detailed calculations, drawings, contracts and so on.

When analyzing concrete business plans as well as templates for business plans at competitions, institutions and from experts in Germany as well as the US e.g. from the competition Start2Grow (Start2Grow 2014) or the government site *BMWi* (BMWi 2014a), or the templates from MIT100k (MIT $100k 2011), the government site SBA (U.S. Small Business Administration n.d.) or Entrepreneur.com (Entrepreneur.com n.d.), there is actually a certain pattern that can be found in each one. Aside from the general three parts already mentioned above, they have even more in common. They usually have seven to ten chapters, sharing at least the following six content areas:

- Executive Summary
- Product & Service
- Market
- Organization & Management
- Marketing
- Financials

---

[4] KPI = Key Performing Indicators

The templates then vary mostly in the way, where they include further aspects like implementation planning, risk and opportunities.

A typical business plan across various nations has about 15 to 20 pages (compare the different sources above), but many investors, for example, are not willing to read so many pages, because they receive usually several plans per day. They rather search for smart teams themselves or wait till their network recommends a startup (Perkins 2007). That's why it is important to have rather a good summary, a PowerPoint presentation with the key aspects, which are quite similar to those listed above, and a business case with the essential calculations in Excel (Fisher n.d.; Romans 2013; German/ US Venture Capitalists 2010). Because of this interest in shorter BP versions, so-called mini-plans or dehydrated business plans have also been introduced (Timmons & Spinelli 2004, p.402). Strange that neither the university institutions and competitions nor the consulting agencies for founders adapt to this generally accepted requirement of investors.

3.13.2 Effectiveness of Business Plans

Many people in the business and investing world as well as some research fellows like William Bygrave from Babson College or Amar Bhide from Columbia University have been examining for quite a while, whether teams without a proper or complete business plan perform better than those having one (Wallstreet Journal 2007; Komisar 2010; Lister 2011). They believe that if a startup is not seeking for external money, no formal business plan is needed. Bhide (1994, p.152) had already conducted a survey in the mid 1990s among America's Fortune 500 companies and showed that 41% did not have a business plan at all.

A general critique – also from the above-mentioned sources – about a business plan is that it often is too formal and distracts teams from the crucial aspects of their business. These considerations can be very different. For some teams, it is more an internal problem, because a certain expertise is missing, while others have to understand their difficult competition or need to understand, what exactly customers want. Successful modern companies like Apple, Google, Facebook or Sun Microsystem have failed with their first BP. According to Komisar (2010), this is called "Plan A". The companies needed to rebuild, plan and test within the markets, what was attractive for the customers. Interestingly enough, they ended up not listening or working with their original plan but were lead to success by the instincts of their founder and CEO. The same reality was told by the German Qiagen founder, Metin Colpan. He said that they had to switch and adapt the company's business setup and thus business plan a dozen time and still the business plan was not working and rather seen as a bureaucratic act.

Their first private investor simply believed in him and his team, not the business plan (Colpan 2008).

Most business plan competitions for startups, however, have not questioned this problem yet or have not found a better way of comparing several businesses and team with each other. MIT's 100k Competition makes a slight exception, by trying to break down the classic business plan requirements. For the first round, only a 2-page summary and three pages for a freelance use are required. In the second round, a full BP has to be handed in, plus a 12-page Power Point presentation. Aside from that, there are no further requirements, how the documents need to be structured (MIT $100k 2011).

Generally speaking, BP competition lead to a problem, because young teams tend to adapt their BP and then also adapt their business focus to the criteria of the competition, instead of following the instincts, beliefs and original motivation, with which they have started with. Despite the positive training aspect, many teams very often fall into a sort of hole, once the competition is over, not knowing how to proceed, unless they had won a price and had some coaching (Feedback from several teams at the CIE in Karlsruhe). The plans too often subliminal suggest that everything is covered and without motivating the team to become more active.

3.13.3 Business Plans in a New Context

Timmons was one of the first to see the business planning in context of an entrepreneurial process (see on the following page figure 9). Unlike conventional business plan models, he first does not want the entrepreneur to start with a business plan, but with a *market opportunity* – a concept almost every investor is preaching to young entrepreneurs. The two other major elements in his entrepreneurial process are the resources and the team. After these three elements met, the BP will follow with the identification of a great opportunity, and thus become less important and disturbing for the entrepreneurs (Timmons et al. 2004). Timmons actually puts the BP on a level like leadership and creativity in his model, indicating that a BP has rather a tool character and is a means to an end. This model fits well with Professor Stevenson's definition mentioned in chapter 3.4, with the aspect of seizing opportunity beyond the resources the founders or startups currently control.

In addition, the author was examining, how entrepreneurs had been developing their business ideas. In over 200 business cases that had been described and presented to him from 2008-2011, one thing that all potential founders had in common was the idea of something new or different, combined with the motivation to do something on their own. These are key aspects and should always stand in the center of the business development.

Since a business plan should not be the center piece of a startup's development, but is necessary for many stakeholders, including the startup itself, to name the required steps to be taken, it becomes a question of timing when to write and improve a business plan and what kind of structure and detail level is needed at each stage.

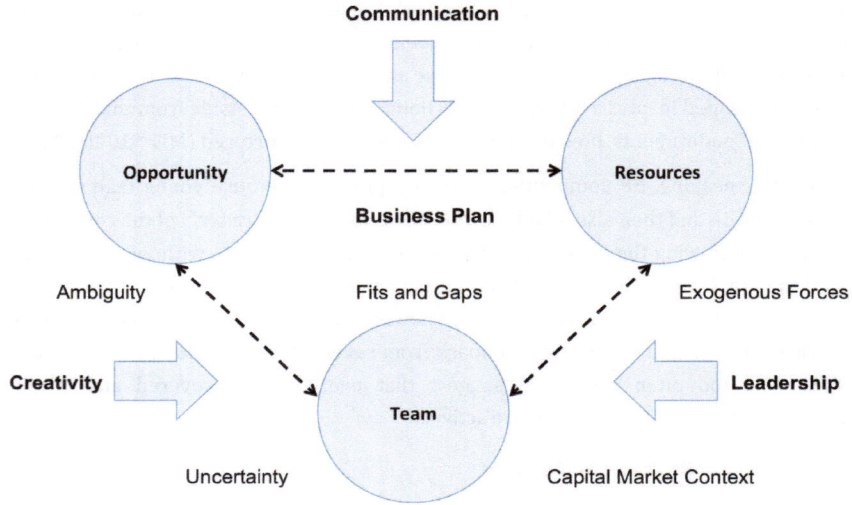

Figure 9: Timmons Model of Entrepreneurial Process (Timmons et al. 2004, p.16)

3.14 Business Case

Business cases are used in any type of company – whether new or established – to describe, quantify and evaluate a business situation or business scenario. A business case shall help to make a decision on an investment, project or initiative (Brugger 2009). It further needs to point out the goals and necessary actions to be taken in order to achieve the (Gambles 2009, p.165). The definitions vary from a pure quantified Excel sheet to a comprehensive document almost similar to a business plan itself. However, the business case – as the name already reveals – focuses on a particular case and it is build-up quantitatively, whereas the business plan includes the whole company from structure, to reports and to calculations (Brugger 2009, p.30).

3.14.1 Business Cases in an Entrepreneurial Environment

In a venture capital or startup environment the term business case is often used equivalent to the calculated business plan version in an excel model (according to various interviews with VCs). This makes perfectly sense, because a startup usually starts with one business model and has one business case for it. Thus, calculating this case means providing the financial and number base for the business plan and the connected decisions. Often investors also call it the financial model or spreadsheet (Romans 2013, p.106). For startups, business cases are usually used as the later financial plan and, therefore, continuously altered from targeted to actual data. The significance from the funding to the operational phase is therefore very high and modeling a business case very complex. For new businesses, the time horizon lies between 3-5 years (Braxton Finance 2013), with a detailed observation on the first 3 years. For established businesses these time frames can be extended up to 5-10 years, depending on the scope and investment of the project or case (Brugger 2009, p.12). The less established a business is, the less accurate the financial data will be. Whereas mature companies and business models can rely on their past and proven numbers, startups on the other hand need to show the potential of their business and the supporting key financial numbers. Thus, experienced investors for example can better judge, whether the estimations are feasible and correct. Besides calculations, a comprehensive business case includes KPIs, different scenario levels, risk observation, as well as key financial data, such as ROI and NPV[5] (Melendez 2008; Schedlbauer 2011, p.35). The KPIs can help steer a project or case by concentrating their attention on the major influencing factors. For a startup, this is also very relevant in the planning process, because it shows the team, as well as the investors, which levers will have the main influences upon the startup's success.

3.14.2 Scenario Planning

In order to picture different outcomes of a business, the calculation of a case often makes use of scenarios. This so-called scenario planning, used by many companies and enhanced by many management consultancies, was first mentioned by American futurist Herman Kahn, as well as by a French group around Gaston Berger (Wilkenson, & Kupers, 2013). It was during the time of game theory and war scenario planning in the early Cold War times of the 1950s (Burton 2011). The oil company Royal Dutch Shell was then said to be the first corporation to have professionally used scenario planning in the 1960s.

The base and most realistic scenario of the business case later can be turned into the financial planning model, which leaves open the option to continue different scenario outcomes. That is why a business case and a business plan are two matching tools for structuring, projecting, planning and communicating the business.

Today, businesses use scenarios to expand their thinking, reveal elements of the inevitable future and avoid group thinking through numbers and facts (Roxbergh &

---

[5] ROI = Return on Investment, NPV = Net Present Value

McKinsey 2009). The base for scenario planning are the past revenues, costs and developments of the existing business, which will be projected into the future, using in- and decreases values from these numbers. Startups, which do not have a past record, need to use industry standards or numbers from similar businesses to project their potential developments.

3.15 Government Support

There are various forms of government support for startups and its investors. One needs to distinguish between direct an indirect support. Direct support occurs through scholarships or special low interest rates for founders and for young companies in general. Indirect support for startups arises, when universities or institutions give funding for the support of startups and their environment. Federal, state and municipal government programs will be further explained in chapter 6.7.8 on government programs within the Entrepreneurial Ecosystem. The support varies of course, depending upon the economic system, which either stands for a very supportive society like Germany or a capitalistic one like the USA, which does not foresee much government support for individuals.

3.16 Competitions

An increasingly important role over the last years have been business plan competitions, that encourage students and non-students to develop business ideas and present them in front of university or business juries. These competitions play an active and important role in fostering the awareness of startups for real business questions. Posters, media coverage, participating companies and information communicated by the competing potential founders spread the word and call attention to entrepreneurship. The website Bizplancompetitions listed 242 business plan competitions in the US in early 2014 (BPC 2014). Of these, startup hotspots like California, New York and Massachusetts already account for more than 50 events. The German website Foerderland.de lists 79 competitions, however out of these 79, there are also awards like the "entrepreneur of the year", which in total decreases the number of actual competitions (Foerderland 2014). Although business plan competitions stimulate potential founders, they should not be over estimated and rather seen as a training module (compare the previous section on business plans). It is dangerous, if teams and aspiring entrepreneurs take competitions too serious and adapt too much of their business model and documents to the rules of each competition. Thus, they will often lose their focus and tend to follow the guideline of the competition, rather than their own intuitive judgment. This is especially the case, if the competition has a panel not run by experienced entrepreneurs. The advice given by these supporters is often too far removed from the reality of a startup or an entrepreneur, and thus might mislead or discourage the potential founders.

Nevertheless there are also many positive aspects of such competitions. Among these are

- Getting things done within a certain time frame, plus at a good detail level of information and moving forward to a rapid pace;
- Good training on how to present the company and receiving an at least partly helpful feedback ;
- Important contacts – networking – for further development (investors, clients, co-workers, mentors);
- Prize money and support.

Eventually, a competition can accelerate and move a team for further and real competition in the market. It should be noted that based upon the author's experience having participated in more than a 20 competitions, as a competitor, jury or audience member, hardly any female participants competed in those events. Whereas in real life one could argue that many women rather prefer a safe job than going through the uncertainty of a startup life. This observation however, cannot be the major explanation for the lack of females participating in competition, while they are still going to college or university. Other reasons may be the lack of aggression and less expressed desire to constantly compete among women. In contrast, the affection for competition among males – whether it is in sports, playing games or in business – is very high. Research by Gneezy and Rustichini from the University of Chicago or Niederle from Stanford University are supporting to an extent some of theses explanations (Gneezy 2002). In several other observations and experiments between males and females of different age-groups between children hood to adult hood, the men tended to have a higher affection for competition, as soon as the situation allowed competition e.g. racing against each other or incentivized problem solving tests. These competitive gender aspects need to be considered later when one is dealing with defining and improving the Entrepreneurial Ecosystem.

4   Success Factors

For an improvement of the Entrepreneurial Ecosystem and the fostering of innovation, it is important to analyze, what it is that makes startups successful. Thus, the ecosystem can be adapted towards increasing the success chances of the business ventures and thus the chances for innovation will increase.

4.1   Success

Looking at the origin of the word success coming from the Latin word *successus,* Merriam Webster (2010c) defines it as *"the degree or measure of succeeding"* or *"favorable or desired outcome"*. Additionally, WordNet defines success as *"an event that accomplishes its intended purpose"* (WordNetSearch 2010). It is obvious that success and its meaning depend on the set of goals or the degree of measurement. The degree of success is a form of desired and achieved performance, and thus can be used as a synonym for success. In other studies, performance has been used to measure venture survival, profits and growth on the base of a subjective assessment (Cooper et al. 1994, p.373).

When an investor is asked to rate his most successful investment, he would most likely choose the one, which best fulfilled his set of goals. In most cases this would mean, the highest monetary value for the investor. One could also assume investments that have had the highest ratio between exit and invested capital. Either way, many investors see success as a financial achievement, which is purpose of the nature of their business – investing money.

From the entrepreneur's perspective, the definition of success must be multi-dimensional. On the one hand, entrepreneurs with primarily monetary goals have proved to have better odds of achieving and succeeding (Gurdon & Samsom 2010, p.207). It is, therefore, not surprising that financial goals and success are also crucial for them. Still, entrepreneurs do also have personal goals, like creating something new and working independently, having employees and living their long-standing dream. Their dream broadens their view of success and makes it also very personal. Unless the entrepreneur sets goals beyond financial aspects, success will only be measured on financial considerations. An issue that led, however, to much controversy in the past and will lead to more in the future, since the money focus is too one-sided, leaving out important aspects such as social, personal and business life. The results of such one-sidedness, is more government restrictions, more protective rights of employees, more labor unions strikes and negative media coverage. Thus, entrepreneurs should take the opportunity and hold themselves back from a pure money focused business style, and instead encourage themselves to be responsible, sustainable and meaningful individuals.

The definition of success should therefore be extended beyond a pure monetary consideration and also include other values like ethics as well as quality of life. This could be motivated both, from a moral and from a business perspective, in which contented employees with social security in end could lead to even more profitability, a more sustainable business and less government interaction. That is why success cannot only be considered for the individual entrepreneur but also for the connected employees and ultimately for society. This expanded dimension indicates an entrepreneur's success also includes his or her environment. The ecosystem will then also need to be challenged from different perspectives, in order to evaluate whether it requires further changes to fulfill this expanded definition of success. Innovation is a good example for it, because its application in the market will not only bring personal success to the entrepreneur, but make the advantages of the innovation available to the public.

A first movement into this direction can be seen through the rise of social entrepreneurship (The Economist 2006). There has also been a doubling of donations in the past 20 years in the US for example from 160bn to 320bn USD and the increasing desire of wealthy individuals to give back to society (National Philantropic Trust 2013). Many well to do persons are citing Andrew Carnegie, who has said "*The man who dies rich dies disgraced*", and thus one can speak of a positive trend also impacting the success definition (Public Broadcasting Service n.d.).

4.2   Characteristics of Entrepreneurs

Throughout decades experts and researchers have been pondering about the characteristics of entrepreneurs and about those that make the difference between success and failure. Timmons & Spinelli (2004) have come up with six general characteristics of entrepreneurs that can also be found in other studies:

1) Commitment and determination
2) Leadership
3) Opportunity obsession
4) Tolerance of risk, ambiguity and uncertainty
5) Creativity, self-reliance and ability to adapt
6) Motivation to excel

Except for leadership, five of these six characteristics of entrepreneurs could easily be used for identifying someone, who is best suited for a solo activity that entails little or no interaction with others, such as racing a sailboat alone. The emphasis on individualism is so dominant in western culture that many of the entrepreneur's characteristics hardly reflect that launching a startup involves constant social interaction with others (Byers et al. 1997). The question, whether these individual characteristics should rather be enhanced by a social interaction view, must be answered in the affirmative, since throughout an entrepreneur's career, there will be an interaction with customers, employees or partners that can make the difference. Doing business is always dealing with people successfully.

Total commitment, opportunity obsession and motivation to excel appear to be a description of the kind of harsh individualist, who struggles alone to win under any circumstance (Byers et al. 1997). Only leadership of the six characteristics refers to a social nature of entrepreneurship, which the individualist generally is lacking.

Aldrich & Zimmer (1986) propose that entrepreneurship has to be seen in a social context, channeled and facilitated or inhibited by a person's position in a social network. These social networks facilitate the activities of potential entrepreneurs by presenting them opportunities as well as providing them with resources for a new venture. This suggests that success does not depend just on the initial individual situation of the entrepreneur, but also on the personal contacts he establishes, maintains and extends throughout the process (Byers et al. 1997).

Therefore, entrepreneurs must be especially skilled at using their time to develop relationships with people who are crucial to the success of their new venture (Freeman 1996). A new venture may start as an idea or hobby of one or more persons, but it takes many more to bring it to life, to make it larger and to grow into a successful company. The first steps usually can be taken with the existing network of the entrepreneur – being "insiders": friends, family and co-founders. As the foundation of the economic venture increases, however, startups need to reach beyond their individual social network and involve "outsiders" such as banks, venture capitalists, lawyers, accountants, strategic partners, customers, and industry analysts and influencers, which bring in their own networks (Byers et al. 1997).

Howard Stevenson (Stevenson 2000, p.1) former dean and professor at Harvard University came up with some hypotheses in 2010, after having examined history and culture in more than 40 countries over the last two decades, which leads to success of entrepreneurship in a broader way:

- Entrepreneurship flourishes in areas with mobile resources.
- Entrepreneurship is larger when successful entrepreneurs and businesspersons reinvest capital in the ventures of other members of this area
- Entrepreneurship flourishes in areas in which success is celebrated.
- Entrepreneurship is better in areas that understand change as something positive and not negative.

The observations of Stevenson indicate further characteristics of entrepreneurs and success factors. Entrepreneurs need to be open for change and move towards the opportunity and the resources, when interpreting Stevenson's first observation. The second leads to sort of cycle, when successful entrepreneurs help the next generation of entrepreneurs to become successful, especially by investing their capital. The last two observations are more focused on cultural aspects of society, however also the entrepreneurs need to embrace theses values.

Burgstone and Murphy (2012, p.9) have researched how entrepreneurs should structure themselves in order to be successful and have identified seven components:

1) Finding and filling an unmet customer need
2) Planning profitability
3) Striving for sustainability
4) Establishing credibility
5) Gathering necessary resources
6) Leading and managing effectively
7) Maintaining balance and learning to enjoy the ride

In their book "Breakthrough Entrepreneurship" Burgstone and Murphy mentioned also in the chapter of gathering the right resources, the three most important resources for an entrepreneurial endeavor are time, people and money – an insight the Google founders Page and Brin experienced on the endeavor of financing and building up Google (Burgstone & Murphy 2012, p.156). These three factors could also stand for crucial success factors and at the same time they show, how many different success factors exist depending on the perspective of the entrepreneur, the investors or society and on the level of detail.

4.3   Entrepreneurs as Social Creatures

Social characteristics of an entrepreneur play an important role in their success, along with knowledge and other skills. This aspect is supported and described in the book "Fresh Perspectives: Entrepreneurship" by Co & Groenewald (2006, p.360), in which they stress out that entrepreneurship is a social process and that the entrepreneurs are acting more social than solo. However, Co & Groenewald also mention that this should not lead to the conclusion that the entrepreneurs' skills and expertise are irrelevant for their companies' success. They rather suggest that research and teaching of entrepreneurship should also focus on social behavior, in order to strengthen the entrepreneurs in dealing with internal and external relationships, to help them in identifying the essential relationships for their success and teach them how to maintain these important ties. Ultimately, they must also learn how to acquire the necessary resources – from the right information to the ideal fund – for their company to succeed.

Entrepreneurship should be seen as a social activity and thus aspiring founders would benefit from getting to know the basic tools of social network analysis (Burt et al. 2013). This knowledge could be used for the understanding and identifying the right connections between key people and companies as well as which players are fundamental for a startup's success. Freeman (1996) relates this aspect to the contacts and the network VC companies provide to a startup and points out that these social aspects might be more important than the money they provide. Due to the importance of these contacts and the knowledge, as well as experience the investors can provide, one speaks today often of *smart capital* if besides the money these extra values are available. Freeman (1996) quotes the well-known Mayfield Fund, which claims it has close ties to technology leaders, academic institutions, other VCs, financial institutions, consultants and firms throughout the US and the globe. They believe their contacts are a key resource for a growing technology-based company For a startup it is therefore important to consider these extra benefits and contacts when seeking funding.

According to Professor Dorf (1998) of the University of California, if an entrepreneur can determine the crucial contacts for his or her success, then most of his or her time could then be spent wisely on building and maintaining these connections. Like any leader, entrepreneurs also need to be strong in convincing and in persuading – employees as well as customers. Many famous entrepreneurs and business leaders are notorious for their interpersonal skills. Dorf mentions Herb Kelleher to be a good example. Kelleher is co-founder and CEO of Southwest Airlines, one of the world's most successful airlines. Kelleher is said to have been a master of persuasion. He can be charming, flattering or joking with employees as well as with customers. And he is also known to bully his competitors and strongly argue with politicians. Not all entrepreneurs and business leaders might be born with this gift or ability, however parts of it can be learned as a skill, and therefore the tools can be researched and be taught as interpersonal influence courses in institutions around the word (Dorf 1998). Chapter 6.4 about the Entrepreneurial Ecosystem will discuss these skills and abilities in detail.

The final persuasion of the entrepreneur often is the result of smart negotiations, which occur with employees, investors, strategic partners, suppliers or customers. Many of these parties are also mentioned in the Business Model Canvas, which today is quite often used by startups and companies to understand and to define their business model (Osterwalder et al. 2010). The negotiating skills ads an interesting component to this model, because understanding and defining is one thing, changing and making it happen another. These social interactions, however, are so complex that further research is needed, to better understand how the skills help in negotiating, persuading and convincing key persons, and thus are an essential skills set of an entrepreneur.

Nevertheless, successful startups as well as prospering established companies always consist of a whole group of people, who make it happen. Therefore, the entrepreneur needs to be in permanent contact with his or her employees, partners, shareholders etc. Most of the time there will be interaction with all stakeholders, on which the success of the companies in the end depends upon. One can often hear managers complaining about one meeting following another leaving now time for real work. But the necessity of many of those meetings rather calls for a skillful training how to run such meetings more effectively. Managers, as well as entrepreneurs should acquire skills like group dynamics, leadership or creativity techniques for being more successful with their projects and ventures (Eisenhardt & Schoonhoven 1990). In many cases, a company's and entrepreneur's success comes down to the team performance and their success (Reich 1987).

4.4   Important Factors for Success of Startups

There are many examples of successful and failed companies, which make it at least partially possible, to identify a few factors that have a strong influence on the success for a startup. Experts mention that most startups have not failed because of a missing business potential, but rather of a bad and only short focused realization concept (Kollmann 2006).

It can generally be said that companies, in which VCs have invested in, have a greater chance of survival. Thus, it makes sense to consider the aspects VCs look at in their decision and monitoring process, because they could also be success factors of a startup. According to various investors like Romans, Manger or Überla (all cited in this thesis), there are in general three areas, which should be considered, when talking with investors about startups and their success chances. The first are the team, the second the product and third the market, which must be dealt with, aside from those mentioned in part by Timmons and by entrepreur.com in the business plan section 3.13 of this thesis.

However, Professor Jacobsen for example has observed in her thesis on success in entrepreneurship that aside from the many studies and research projects that have occurred, their comparability and consistency is lacking (Jacobsen 2003, p.18). Upon a closer examination, one will realize that startups, entrepreneurs and their environment are or can be so different that these comparisons tend to be very complex and almost impossible to rely upon, thus making a stable academic model for successful startups and entrepreneurs quite, if not very difficult to develop and use fully.

4.4.1   Entrepreneurial Team

The team factor is the number one issue to deal with in discussions about startups and their success. Whether one talks with dozens of investors or one studies various papers about success factors of new ventures in the end, it always gets down to the team (Romans 2013, p.44). A common saying by many investors is: " I have three key aspects to look upon, the team, the team and the team". One also finds part of the allegory of the jockey and the horse: *"There is no question that irrespective of the horse (product), horse race (market), or odds (financial criteria), it is the jockey (entrepreneur), who fundamentally determines whether the venture capitalist will place a bet at all."* (MacMillan et al. 1985, pp.119–128)

It generally is agreed upon that in the end it is the entrepreneur and his team that is determinative in deciding, if the venture receives financing or not. Dubini (1989) found out in his study, which was partially based on the study of MacMillan et al, that also the great attention to detail, the ability to discover and to handle risks were additional decisive indicators of success for new ventures (Brettel 2002, pp.305–320). Other studies, furthermore emphasize the importance of leadership (Zutshi 1999)(MacMillan et al. 1985); the ability of the team to sustain and to put an intense effort into the venture (Sandberg & Hofer 1987) as well as the team's general commitment to the common goals (Chorev & Anderson 2006, pp.162–174). The ability to articulate and describe the business idea was also found to be another important asset (MacMillan et al. 1985, pp.121–122; Brettel 2002, p.311), however it is hard to say, whether these last two mentioned aspects are really so important to success. A study by Sandberg & Hofer (1987, pp.21–22) found out that apart from the initial startup- and entrepreneurial experience, the biographical characteristics like the entrepreneur's age, education and previous industry experience have no correlation with the actual performance of a new venture. On the other hand, they did not exclude the chance that the psychological or behavioral traits might have an influence on the new venture.

A study by Gartner et al (1999, pp.224–235) claims that *"the industry experience may often be a liability rather than a benefit"*. They stated that sometimes past experience leads to sticking to yesterday's rules, which in a dynamic industry may not lead to future success. But when it comes to prior experience from the related industry or the target market, other studies show a positive correlation with it and the success. Riquelme & Richards (1992) found out that a lack of prior experience in a venture's business area was a "no-go" signal for some investors. This stands in direct contrast to a study by Cooper et al (1994, pp.390–391), which declares the industry related know-how is a significant contribution to the venture's growth and survival rate as well as to its success (Shepherd 1999, pp.621–629). Ultimately, prior industry or venture experience can always be helpful, if the entrepreneur has the right attitude and is always open to improve and experiment.

The initial motivation and the aspiration to succeed is also an important factor, which affects the investment decision (Silva 2004, p.130) as well as the focus on financial outcomes (Gurdon & Samsom 2010, p.207). The size of the founding team was also found out to be a factor affecting the ability and the need of the venture to expand (Cooper et al. 1994, p.371). In practice, investors like teams of two to three persons, which a smart mix of skills and abilities, while the dislike single lone warriors or large founding teams with unclear decision streams.

4.4.2   Product and Service

VCs' investment criteria include of course measurements of the venture's product and service. Furthermore, investors are interested in product characteristics like the uniqueness of the product (Silva 2004, p.130); (Tyebjee & Bruno 1984, p.1061) as well as certain IP rights – patentability or protection against imitation (Hall & Hofer 1993, p.27). Investors consider IP protection also as an entry hurdle for following competitors, thus giving their investment a head start. Moreover, the protected technology can be always an important element for a later sale to another company. In addition, investors pay much attention on the realization and market position. They prefer an existing prototype (Sandberg & Hofer 1987, p.129; MacMillan et al. 1985, pp.129–130) with a confirmed market acceptance. This makes financing much easier (MacMillan et al. 1987), because in the end it is always the customer, who decides over the success. A superiority of competitor's products is also an important factor for sustainability (Zacharakis & Meyer 1998, p.70; Fried & Hisrich 1994, p.30). Eventually, a potential customization and an functioning strategy for it, is crucial in many dynamic markets and will increase their survival chances (Gartner et al. 1999, pp.225–226). Especially in today's markets, individuality has increased immensely. Thus, customized products are easier to sell and lead to higher better margins than mass market products (Kakati 2003, p.452).

4.4.3   Market Characteristics

Ultimately, the customer will decide over the success of a company. Thus, the market evaluation, which is an important factor, becomes essential. The best team with the greatest product will have problems or will eventually fail, if they are not ready, or if the market dies not exist. Sometimes even an irreproducible product, which is better, newer or superior, fails. Even good relations or connections between customers and other competitors, a complex supply chain with hard access for new products, slow moving customers, as well as ongoing long-term contracts can hinder a market entry (Sandberg & Hofer 1987, pp.5–8). Nokia did not succeed with its 7710 touchscreen in 2004, whereas Apple turned the phone market upside-down with the iPhone and its touch screen (Hage 2007). A mixture of timing, marketing, supplier management and product features can make the difference. When unexpected reactions and developments within the markets become problematic, the investors will often rely on the team, with whom they had their first success with. A good team might not be able to turn a market, but be smart and fast enough to change the product or strategy, to find a new market and adapt further to the dynamics, till the customer need is properly identified and satisfied.

Some studies indicate that it is easier to be successful in a market that it is large and attractive (Tyebjee & Bruno 1984) and still growing at a high rate (MacMillan et al. 1985, pp.120–122; Hall & Hofer 1993). Some products might even stimulate the market (MacMillan et al. 1987; Brettel 2002, pp.310–312). For this, it does not necessarily require a fancy product of Apple to do so, but also very classic companies like the German SAP were able to influence and stimulate the market. When SAP introduced to the business world the new changing R3 software, the customers did not expect or even at first did not want it. Nevertheless, SAP's leaders still believed in the R3 program and went on with this new glorious entrepreneurial leap into the market (Zencke 2010).

Further factors that can be analyzed and evaluated in a market for the startup to enter and for later competitors to consider are:

1) The number of competitors;
2) The product life cycles;
3) The maturity of the industry;
4) The legal and economic barriers of entry;
5) The IP protection;
6) The type and size of customer groups;
7) The buying habits
8) The price sensitivity;
9) The marketing and advertising situation.

Especially in the Internet market, companies can calculate how much it costs to reach a certain number of customers. Through Search Engine Marketing (SEM), as a part of online marketing, companies can very effectively attract potential customers to their website and even with a certain probability turn them into paying customers – the so-called conversion rate. The costs for SEM can be calculated in advance and with a prior proof of visitors actually converting to customers, it is possible to form a ratio between necessary marketing investments and sales. Once a startup can convince an investor to invest in them with the intention of using parts of the financial means for this type of online marketing, a certain rate of success can be well estimated. However, this type of financing and online marketing does not necessarily guarantee long-term success and sustainability.

4.5   Expert Panel Survey

In order to obtain more success factors, and thus later key elements for a successful Entrepreneurial Ecosystem, the available existing secondary information has to be enhanced by specific primary information. Therefore, a written survey has been conducted asking an expert panel of investors, as well as entrepreneurs, what they believe are the success factors of startups. For both panels, the idea has been to collect participants from different countries, bases on different economical size and their level of maturity. In order generate a first list, BAs, general investors and VC funds from different countries from the author's network have been collected. Further relevant investors from the German and US venture scene (investors, BAs, professors, entrepreneurs, journalists and startups advisers) – also from the author's private network – supplemented the list. In order to qualify for the investor's circle, only investors with an active record of at least 3 investments have been considered. In 2011, 50 investors from the list were contacted, 28 mainly from Germany, the US and Finland participated in this final expert panel. Parallel to this panel, a list of entrepreneurs from the author's network, as well as from the venture scene was established and studied. They ranged from first time founders to successful serial entrepreneurs. In order to qualify for the entrepreneur's pool, founders must have gone at least once beyond the founding process, running or having run a registered company. This way, 240 entrepreneurs were contacted, of which the final 40 participants – completing the entrepreneur expert panel – came mainly from the US, Germany and the Netherlands. In order to compare the data from both participating groups, a common setup and standard for both surveys had been chosen, based on the results from the existing research.

The general setup included the three areas described on success in section 4.4 plus financial aspects as mentioned in the business plan section 3.13:

1) Product/Service Characteristics
2) Market Characteristics
3) Venture Team Characteristics
4) Financial Characteristics

In addition, the setup also allowed to mention other factors and to give summarizing thoughts, as well as leaving the participants space for not anticipated answers. The author has used a structure and survey technique applied in consulting business (e.g. inspired by Siemens Management Consulting, Bain & Company, Business Net Partners). After the selection of the participants, the surveys were conducted online in an anonymous form, especially because of many delicate and personal questions. Monetary numbers had been answered in US Dollars or Euros. The intention was to receive input for a deeper understanding of success factors and different perception forms from the investors' and entrepreneurs' perspectives. The results are part of the modeling process of an Entrepreneurial Ecosystem. They have no statistic relevance in terms of size, independence and quality level for assumptions.

4.6   The Investor's Perspective

The complete questions and answers can be found in the Appendix A. In the following are some of the most interesting answers, which will be highlighted and evaluated.

4.6.1   General Aspects on the Investors' Most Successful Companies

The investors were asked to think about their 3-5 most financially successful investments and answer to the questions according to these investments. The following chart shows in which industries these 3-5 most successful companies were operating.

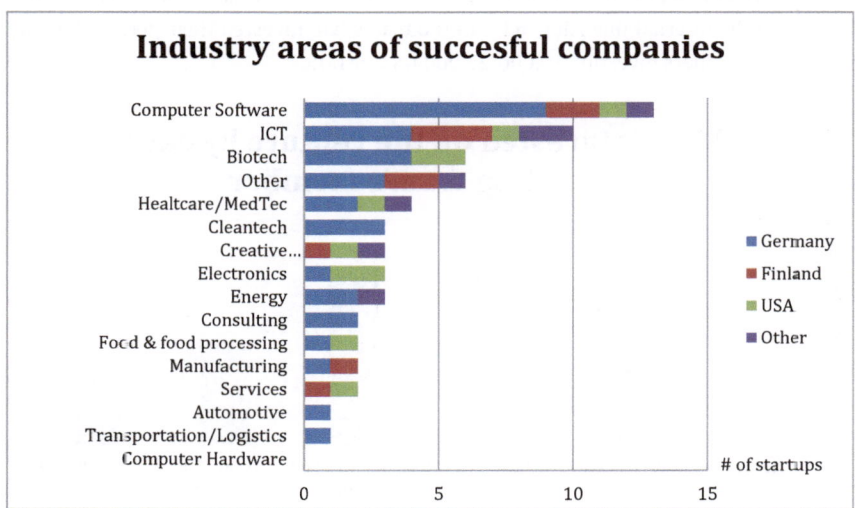

Figure 10: Industry Areas of Successful Companies (Investors could choose several categories)

The most successful investments came from computer software industry, with 46% of the 28 investors mentioning it. The Information and communication technology industry, with 36% and biotechnology with 21%, were the next successful startup groups. The "Other" category consisted of six further industries, which were coating, mobile and wireless, mobile telecom, E-Commerce, Internet and Internet services. When integrating five of these six to the ICT industry, it leads to the biggest industry with 54% of the successful investments.

Surprisingly, the investors did not have a successful venture in computer hardware, although 8 investors had that industry in their portfolio. The success rate for the "classic" German industries, Automotive and Manufacturing, was quite low, but as expected, since these are not typical early venture domains – especially for startups. Cleantech with 10% and Healthcare/Medical Tech 18% were also not among the top listed. This is probably due to the high investments and risks involved, causing to fewer companies to invest in these industries, and thus on the other hand leading to fewer success cases.

When considering the team size, 59% had a founding team that consisted of 1-4 members, while 39% had a team size of 2-3 members, standing for the average size of a founding team (Appendix A, question 5)

The investments had mainly been made in the Seed (35%) and Startup phase (40%). Regarding the internal financing, 63% of the teams had invested their personal money with most members investing 10,000-25,000 USD or EUR.

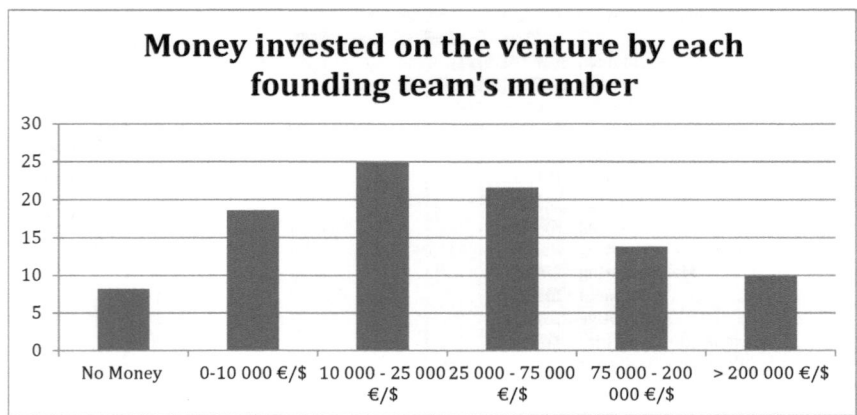

Figure 11: Money Invested on the Venture by the Founding Team's Member

4.6.2   Product and Service Characteristics

With the following question, investors were able rate products and service characteristics with points from 0 (lowest) to 100 (highest). Competitive advantage of a product was given the highest value by the investors with an average of 84 points. Similar high value with 83 points was the fulfillment of market needs. The following two factors also had a little gap to the former mentioned with "There was a functioning prototype of the product (71 points) and "Product uniqueness" (70 points). Protected IP (patents, complexity)" with 63 points and "Customizable products/ services to the customer's needs" with 53 points came in last, however the standard deviation (SD) was higher, indicating that it played for a few investors a much more important role, while others did not value it too high. This was expected, because of the different portfolios, one, which had Internet companies with usually no protected IP, the other had high-tech firms with IP rights in it.

Answer	Average Value	SD
Product/ Service had a competitive advantage	84 pt.	22
Product/ Service would fulfill market's needs	83 pt.	23
There was a functioning prototype of the product	71 pt.	28
Product/ Service was unique compared to the competitors	70 pt.	25
Product/ Service had protected IP (patents, complexity)	63 pt.	30
Product/ Service was customizable to the customer's needs	53 pt.	34

Table 12: Product/ Service Characteristics (Scale from 0-100pt, n=28 (investors))

Fried & Hisrich (1994, pp.28–37) had also listed "Competitive advantage" as one of the fifteen criteria, investors pay a special attention to, when screening a company.

Uniqueness of a product was said to be the most important factor for the expected return on investment according to a study by Tyebjee & Bruno (1984). This uniqueness factor is confirmed by the results from this survey.

When interpreting why IP rights have not been voted to be the decisive element for a venture's financial success, one has to keep in mind another aspect beside the different portfolios. Mainly that a great team can have success in a market without any IP protection such as patents, on the other hand, protected IP alone is worth nothing, unless someone starts to use the leverage with it. IP – besides patents also for example protected brand rights – also starts playing a role much later in a startup's life, after the market has already been entered. In many cases startups do not even reach that point, because they lack initial funding, cannot find customers or fail to get their product of and running. Universities, for example push very hard to attain more patents and protect them. They try to use their leverage by selling licenses to large corporations, while they face great challenges, when the patents should be used in a startup. In most cases, the difficulty comes from the fact that the researchers and scientists are more inventors than entrepreneurs. And teaming them up with entrepreneurs is quite challenging for a university. Thus, they often leave the usage of IP rights to existing companies. That is why IP is a great add-on for a startup but always comes after the team for example that can make use of it or after the growth potential of a market becomes a reality.

4.6.3   Market Characteristics

For the investors the high growth potential of a market was by far the most important aspect, given 80 points on average. The "There was a demonstrated market acceptance of the product" answer followed behind with 55 points, as well as "The venture had already the first customers" with 52 points. For startups this might be a little surprisingly, because it always seems that investors, at least in Germany are looking very much more for the market acceptance and the first customer base. The reason for this is that investors in general do not like to invest in the market, where potential is too small. After that, they pay attention to the market acceptance and the first customers. It does not work well the other way around, because a high market acceptance or first customers in a low potential market, is for most funds uninteresting. What investors were not really afraid of, was the factor "There was a strong competition on the market" getting only 41 points on average.

Answer	Average Value	SD
Market had a high growth potential	80 pt.	27
There was a demonstrated market acceptance for the product	55 pt.	25
The venture had already the first customers	52 pt.	26
The venture stimulated existing market	52 pt.	25
Market was already big	45 pt.	21
Market was new	45 pt.	34
The venture had established distribution channels	42 pt.	31
There was a strong competition on the market	41 pt.	34

Table 13: Market Characteristics (Scale from 0-100pt, n=28 (investors))

4.6.4   Venture Team Characteristics

"Team had understanding of their prospective customers" and "Team members were able to present the business idea well" was given 76 points each, and thus the most important characteristics concerning the team. The latter factor was also found to be important in studies by Macmillan et al. (1987), Dubini (1989) and Brettel (2002). "Capability of sustaining intense effort/durability" was quasi as important with 75 points. Macmillan et al. (1985) have also identified this to be the crucial factor in their study.

Answer	Average Value	SD
Team members were able to present the business idea well	76 pt.	26
Team had understanding of their prospective customers	76 pt.	27
Team was capable of sustaining intense effort/durability	75 pt.	24
Team could react well to altering circumstances	69 pt.	21
Team had experience from the target market	68 pt.	30
Team was capable evaluating and reacting to potential risks	66 pt.	30
The business had an aggressive growth model	66 pt.	30
The majority of the team members had a great attention to	63 pt.	26
Team had proven leadership skills	60 pt.	24
Team was referred by a trustworthy source	56 pt.	27
Team members had different backgrounds (studies, ethnicity, nationality...]	55 pt.	30

Table 14: Venture Team Characteristics (Scale from 0-100%, n=28 (investors)*)*

In a following sub-section, more information on the venture team had been requested concerning their functional expertise. It came out that the most successful teams were experts in "Technical skills/engineering and services" according to the investors, which valued these skills with 83 points. Expertise in "Sales/Marketing/Distribution" followed far behind with an average of 55 points.

Answer	Average Value	SD
Technical skills/engineering/services	83 pt.	17
Sales/marketing/distribution	55 pt.	28
Organization/ administration	52 pt.	27
Finance/controlling	50 pt.	27

Table 15: Functional Expertise of the Team (n=28 (investors))

This high valuation of technical expertise is widely spread internationally and can also be observed in the venture capital scene. Investors tend to look for teams that have the skills to implement important technical parts of their business themselves – at least in the beginning, and ideally are able to distinguish their product from its competitors. The other aspects have shown some variation, with Finnish investors valuing expertise in *"Organization/Administration"* only with 36 points, while Americans gave it 64 points. Due to the small expert panel of these, a separation of the investors, into their countries' origin, does not allow a general statement, whether these positions of American or Finnish investors are representative for their countries.

Before giving the investors a table with choices about important factors and characteristics, they were asked to write down up to 5 reasons why startups fail and why in their portfolio some were successful. The 28 participants listed 114 reasons (ø4.07 reasons per participant) why startups companies can fail and 118 reasons (ø4.21 reasons per participant) why they are successful. The top answer for both is the 'Team & management 'being mentioned in up to almost 40% of the failures and 50% of the success cases. This correlates very well with the general statement one can hear from many investors at national and international competitions or read on their websites or deduce from private talks. And even the cluster 'Strategy & business model' could be related back to the team, adding over 10% of failure and success reasons to the team. 5% of the success reasons were also related to luck according to the investors, whereas, none of them mentioned bad luck as a reason for failure.

Reasons for failure

Cluster	Frequency	%
Team & management	44	38,6%
Market (size, timing, competition, customers, government)	27	23,7%
Strategy & business model	15	13,2%
Product, technology & IP	11	9,6%
Financials & sales	11	9,6%
Investor relations and their performance	4	3,5%
Bad luck	0	0,0%
Other factors	2	1,8%
Total text answers	**114**	**100%**

Table 16: Reasons to Fail – Mentioned in Free Text Fields (114 answers from 28 participants clustered)

Reasons for success

Cluster	Frequency	%
Team & management	60	50,8%
Market (size, timing, competition, customers, government)	15	12,7%
Strategy & business model	14	11,9%
Product, technology & IP	11	9,3%
Financials & sales	3	2,5%
Investor relations and their performance	3	2,5%
Luck	6	5,1%
Other factors	6	5,1%
Total text answers	**118**	**100 %**

Table 17: **Reasons to Succeed – Mentioned in Free Text Fields** (118 individual answers from 28 participants clustered in the above categories)

What the tables above from the survey show is what has also been identified as the main source of the success or the most important investment criteria many times before (Tyebjee & Bruno 1984); (Tyebjee & Bruno 1981); (Robinson 1987); (Sandberg & Hofer 1987); (Muzyka et al. 1996): The team and the management are the most important success and failure factors. Already the first studies in the 1970s identified the management (J. M. Johnson 1979, pp.31-38), the management commitment (Wells 1974) and the quality of the management (Poindexter 1976) as the most important criteria which investors use to pick new portfolio companies and this study supports the results.

4.6.5 General Factors

In the last section, the investors were asked to rate further general 26 different factors. The alternatives ranged from "Motivation" to "Luck" or "Government support". The factors were rated on a scale from one to six, with one meaning "Not at all important" and six meaning "Extremely Important".

The top four ranked factors are purely team-related: "Motivation" with a value of 5.82, "abilities of the founder" at 5.61, the "team performance" with 5.61 followed by a "clear vision" with 5.32. Support from different sides like government, academic or family was ranked relatively low.

	Statistic	Investor's perspective				Startup's perspective		
		Min	Max	Average Value	SD	Average Value	SD	Rank
1	Determination & Motivation of the founder	5	6	5,82	0.39	5.60	0.88	1
2	Abilities and skills of the founder	4	6	5.61	0.57	5.19	1.14	2
3	Team performance	5	6	5.61	0.5	5.02	1.31	6
4	Clear vision	4	6	5.32	0.72	4.98	1.24	8
5	Quality of your product/ service	2	6	5	0.98	5.04	1.04	5
6	Marketing Strategy	3	6	4.93	0.9	4.26	1.37	11
7	Network of contacts	3	6	4.93	0.9	5.06	1.26	4
8	Product/ service innovations	1	6	4.82	1.06	5.00	1.27	7
9	Feedback from the customers	2	6	4.82	1.06	5.13	0.97	3
10	Investors/ Business Angels	2	6	4.82	1.02	3.04	1.99	22
11	Unique Opportunity	3	6	4.61	0.88	4.43	1.28	10
12	Customer loyalty	3	6	4.61	1.03	4.64	1.15	9
13	Competitive Market	1	6	4.18	1.12	3.68	1.24	15
14	Luck	2	6	3.93	1.09	3.64	1.44	17
15	Business Plan	2	6	3.89	1.1	3.55	1.44	19
16	Infrastructure/ Resources	1	6	3.79	1.13	4.19	1.28	12
17	Mentoring/ Coach	1	6	3.68	1.42	3.60	1.38	18
18	Culture of your region	1	5	3.54	1.17	3.02	1.42	23
19	Sticking to the company's values	2	6	3.39	1.1	4.15	1.32	13
20	Government support/ grants	1	5	3.04	1.35	2.83	1.77	24
21	Academic support	1	5	2.96	1.14	3.17	1.42	21
22	Support from the family	1	5	2.89	1.07	3.68	1.53	16
23	Public visibility of company's values	1	5	2.86	0.97	3.81	1.53	14
24	Support from the banks	1	6	2.64	1.39	2.45	1.61	25
25	Support from the friends	1	4	2.54	1	3.49	1.75	20

Table 18: Ranking of the General Factors (Scale from 1 (not at all important) to 6 (extremely important))

The table also already uses the results from the survey of the entrepreneurs, which will be explained in the upcoming section 4.7, in order to offer a comparison between the investor's- and entrepreneur's view. The comparison shall further help, to identify those success factors, which can be related, at least in parts, to the Entrepreneurial Ecosystem.

In the table above, the differences between the two stakeholders become obvious. The investor, of course, sees these aspects from his or her perspective – based upon experience from several different companies. The entrepreneurs normally only rely upon experience from their own company(s).

Motivation and abilities & skills are for both groups the most important aspects. And Motivation fuels every further action, whether it is research, sales, management or activities the team engages in. Motivation supports focus, which helps the team to concentrate on the right things. If there are obstacles to an action, such as a lack of knowledge, motivation will help the team to acquire knowledge or find alternative ways of achieving the goal. It also ensures the capability of sustaining intense effort, no matter what problems might occur. Robinson (1987) identified personal motivation of the founder as a critical factor for the firm's success.

Abilities & skills, as the second most important factor, stand for all the factors the entrepreneurs either naturally bring to the table or which they can acquire by learning and training (compare Ability & Skills 6.4.1.). By a huge variety of different skills mentioned, it is obvious that they play an important role, especially because they represent the aspects entrepreneurs can control and improve themselves.

The investors rank the team performance as the third most important aspect, whereas entrepreneurs see their customers in this position. This different ranking is quite understandable, because investors look upon the team and see them as the key for their fulfillment, in contrast to the founders, which look at their environment and see the customers as the decisive element, because of the money they bring in. Aside from many more fine differences, both rank business plans at the lower end of this scale, underling its meaning of being rather a tool than an important success factor.

Government grants and academic support are ranked very low. Bank support is for both at the bottom of the 26 general factors, with number 24 and 25 in the ranking. When considering the original role of banks, to provide companies with financial means to build up, grow and conduct business, it is a astonishing to see, how little modern banking finances young companies and thus fails to invest in the future of society. Surprisingly, startups in this survey ranked the importance of BAs and investors only on position number 22. This lack of importance could be explained through a classic entrepreneurial attitude, which states that entrepreneurs are solely independent and responsible for their success and failure. Emphasizing the importance of another person like an investor too high, could take away part of this independence. Since the survey had been conducted anonymously after the selection of the participants, there was so option for interviewing participants for further questions about their answers.

4.7   Success Factors from the Entrepreneurs' Perspective

The survey for the entrepreneurs used the same structure as the one for the investors described in chapter 4.4. 62 entrepreneurs from the panel group filled out only major parts of the survey, while 40 completed all questions.

The entrepreneurs' companies were on average 2.7 years old; they had 2.8 Co-Founders and 6 employees. The founders had founded 1.5 companies before their current venture and are paying themselves a fixed salary of 53,000 € p.a.

4.7.1   Market Characteristics and Strategy

In the first section three basic questions on the market, its situation and the strategy were asked. The first asked, which markets the startups are selling their products or services on. 60% of the 62 respondents are selling their products or services on international markets, compared to 32% of them selling national and 6% regionally.

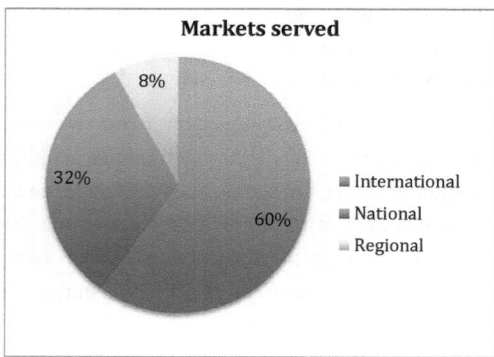

Figure 12: Markets Served by the Startups (n=62 (entrepreneurs))

The second question asked about the market situation of the startups in the beginning of their formation. 40% of the venture teams were competing against big and small competitors, whereas just 3 entrepreneurs said to have had competition from the start with large companies. 26% had few small competitors, while 15% had many. 15% mentioned having had no competitors at all.

The third question was about the strategies of the startups. The following chart shows that "Quality leadership" and "Innovation advantage" where the top ranked with average values of 5.

Answer	Average Value	SD
Quality leadership	5.23	1.51
Innovation advantage	5.02	1.21
Customizable products	4.03	2.09
Cost leadership	3.65	2.17
Strong market penetration	3.63	1.93

Table 19: Startup Strategies (n=62 (entrepreneurs))

4.7.2 Product and Service Characteristics

Investors in the previous chapter had been asked the same questions on product and service characteristics, however instead of giving the high range of points rating from 1-100, the entrepreneurs had only choices from "not applicable", "do not agree", "partially agree" to "totally agree". This survey change was done, because the test group of entrepreneurs, unlike the investors, had problems distributing points.

Almost 75% of the respondents totally agreed that their product has already fulfilled the market needs. 59.68% answered that their product had a competitive advantage. These two factors of market need and competitive advantage were also top ranked among the investors, only with higher values of 84 and 85 points, corresponding to 84% and 85%. The gap is partially coming from the different evaluation scheme mentioned above.

Answer	% Agreement
Your product/ service would fulfill market's needs	74,19 %
Your product/ service had a competitive advantage	59.68 %
Your product/ service was unique compared to the competitors	43.55 %
Your product/ service was customizable	46.77 %
You had a functioning prototype	45.16 %
Your product/ service had protected IP (patents, complexity)	11.29 %

Table 20: Average Product/ Service Characteristic

4.7.3 General Internal and External Factors

The following section were 25 questions concerning the importance to a startup's success. These had been answered by 47 entrepreneurs. This time the rating scheme moved from 1 "not at all important" to 6 "extremely important". This kind of rating was chosen to avoid neutral answers, like in the section above, and to make them again better comparable with the investor's perspective. The highest answer was "Determination & Motivation of the founder" with the high value of 5.6 and a lowest standard variation. It was followed by "Abilities and skills of the founder" and "Feedback from customers". Maybe surprisingly from some investors is the ranking of the "Investors/ Business Angels" with only 3.04 being among the last four valued factors.

Answer	Average Value	SD
Determination & Motivation of the founder	5.60	0.88
Abilities and skills of the founder	5.19	1.14
Feedback from the customers	5.13	0.97
Network of contacts	5.06	1.26
Quality of your product/ service	5.04	1.04
Team performance	5.02	1.31
Product/ service innovations	5.00	1.27
Clear vision	4.98	1.24
Customer loyalty	4.64	1.15
Unique opportunity	4.43	1.28
Marketing strategy	4.26	1.37
Infrastructure/ resources	4.19	1.28
Sticking to your company values	4.15	1.32
Public visibility of your company values	3.81	1.53
Competitive market	3.68	1.24
Support from your family	3.68	1.75
Luck	3.64	1.44
Mentoring/ Coach	3.60	1.38
Business plan	3.55	1.44
Support from your friends	3.49	1.53
Academic support/ Academy	3.17	1.42
Investors/ Business Angels	3.04	1.99
Culture of your region	3.02	1.42
Government support/ grants	2.83	1.77
Support through banks	2.45	1.61

Table 21: **Importance of General Factors** (n=47 (entrepreneurs))

At the end of this section, it was asked, whether internal factors like skills, abilities, the team and the founders' vision are more important than the market situation and its customers. From the 39 respondents this time 4 out of 5 valued the internal factors to be more important than the market (compare Appendix B question 8).

4.7.4   Venture Team Characteristics

53 entrepreneurs have answered the venture team characteristics. The average number of co-founders was 2.7, but when taking out the extreme values of 6, 9 and 14 co-founders, which only occurred once each, the smoothened average comes to a value of 2 co-founders.

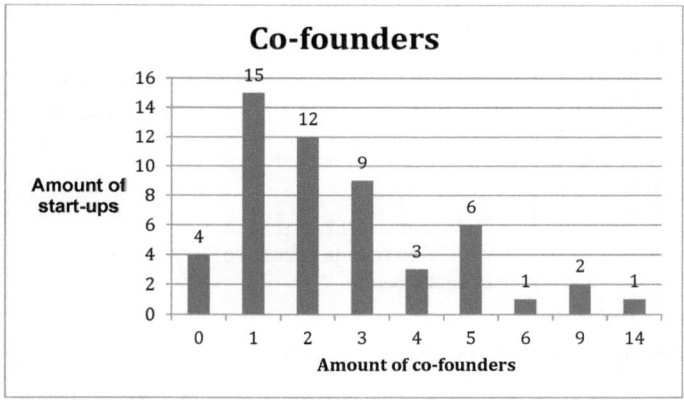

Figure 13: Distribution of Co-Founders (n=53 (entrepreneurs))

When asking about their level of entrepreneurial experience, 50% said to have at least founded one company before.

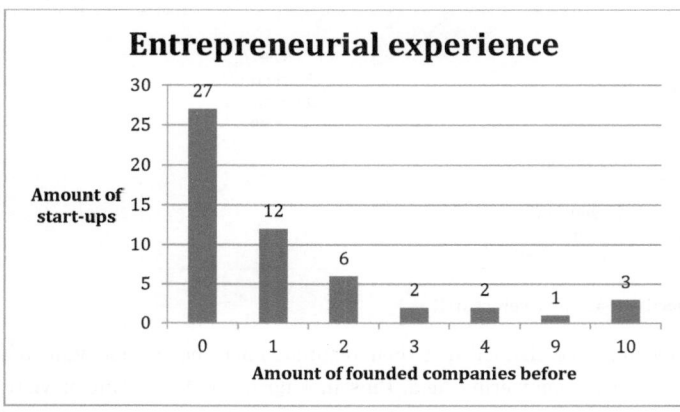

Figure 14: Entrepreneurial Experience (n=53 (entrepreneurs))

It is interesting to see, that the majority of startups indicated to have started their company with between one and four employees. These numbers might have to be further interpreted, because founders sometimes see themselves part of the employee count and reckon free-lancers and part-time workers into it as well.

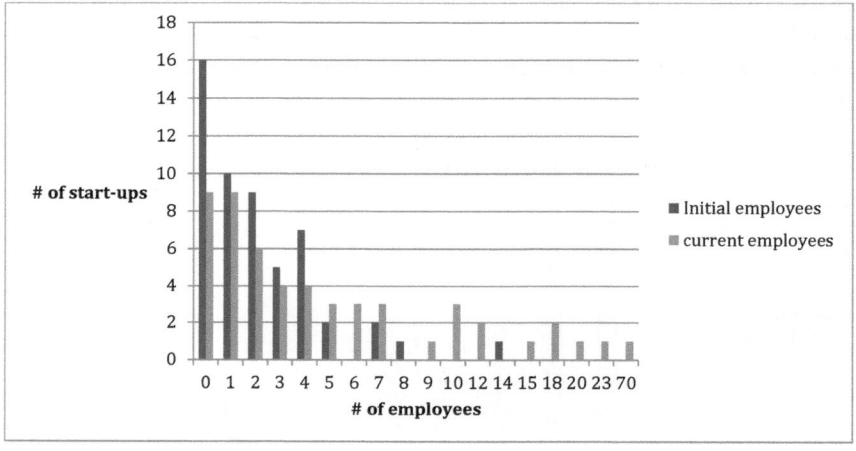

Figure 15: Growth of the Foundation Teams (n=53 (entrepreneurs))

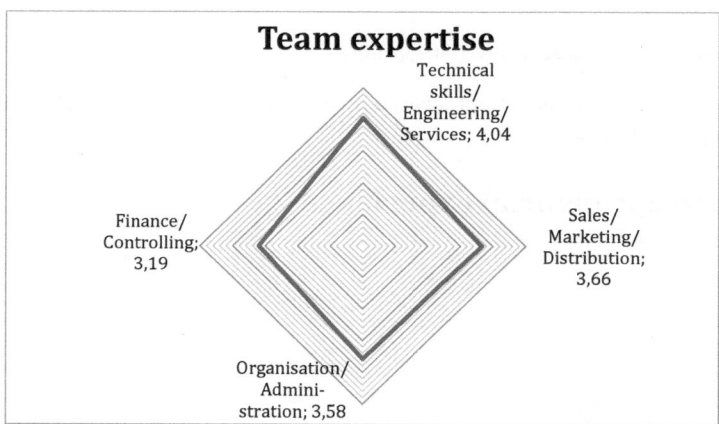

Figure 16: Team Expertise (n=53 (entrepreneurs))

Nearly all entrepreneurs have claimed that their team-members have a medium to a very high expertise in the engineering field, thus making from their point of view, technical skills the most important criteria. However, since the international participants from MIT and Stanford University also have a strong engineering background, the technical importance needs to be considered with caution. Other areas like sales, finance and organizational expertise also play an important role. This is quite understandable, since a startup has to excel in all kinds of business fields, like a decathlon athlete rather than a single discipline contender. When asking them about specific characteristics of their teammates, ability to learn was mentioned to be the most important one, whereas experience on the market came in last. But this also has to be seen in the context of an entrepreneur. Most entrepreneurs are rather young and inexperienced when they start their business. With this inexperience and the fact that they want to start something new, the ability to learn is very important for them. The other team members are usually young as well, because the startup teams-up out of a certain peer group, usually being of similar age. Thus, they all have few or no experience on the market and need to compensate it with the ability to learn. If they were rating marketing experience too high, they would probably never have started a company in the first place.

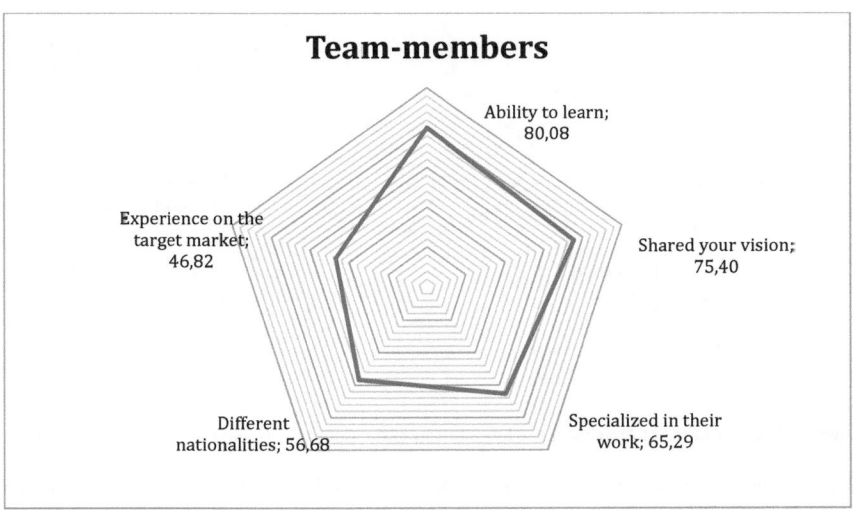

Figure 17: Characteristic of Team-Members (n=53 (entrepreneurs))

4.7.5   Motivation and Characteristics of the Founders

The following section in the survey wanted to know more about the individual team members; respectively the skills, abilities and type of characters. The fist part asked the founders, what has motivated them to start their own business.

The top four ranked aspects with an average value between 68% and 80% are the following:

1) Self-fulfilment (79.33%)
2) Implementation of your own idea (76.98%)
3) Filling of a market gap/ window of opportunity (68.71)
4) Independence (68.13)

Lack of career opportunities or an unemployment situation was ranked at the bottom of the ten possible reasons.

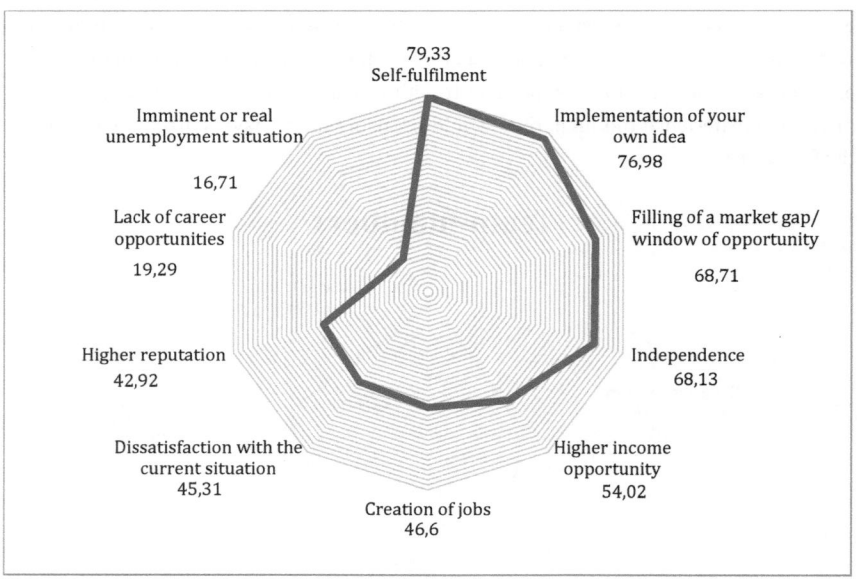

Figure 18: Importance of Aspects at the Foundation of the Company (n=48 (entrepreneurs))

The second part wanted to know more about their skills and abilities. The participants were asked to make a ranking of 10 characteristics. "Creativity" (18) and "Ability to learn" (8) received the most votes for the first places. Together with "Leadership" these three were mentioned the most, when it came to the place one to three. The following chart shows results:

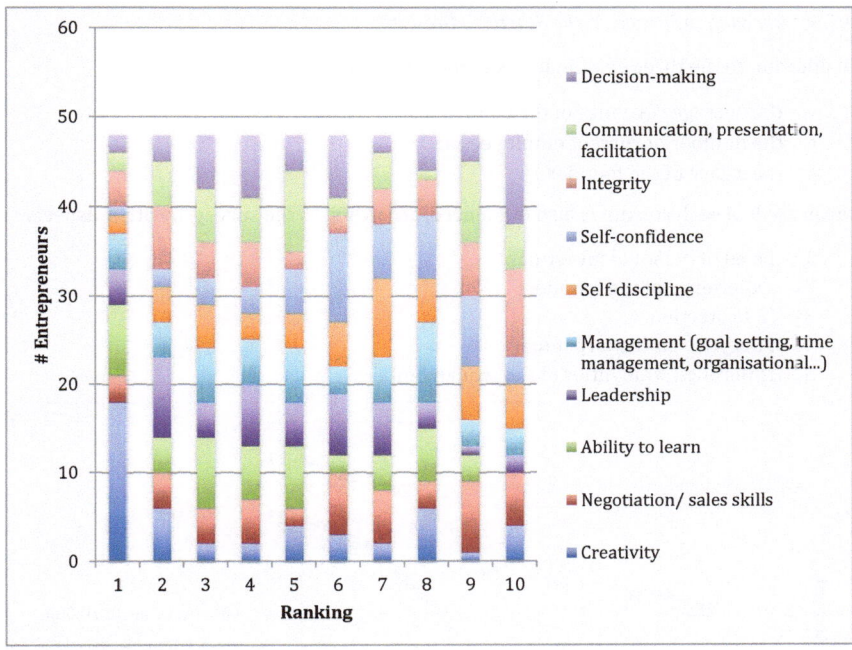

Figure 19: Ranking of Needed Skills and Abilities (n=48 (entrepreneurs))

In addition to the skills and abilities, the entrepreneurs had to evaluate their different attitudes when founding their company on a scale from 1 (very low) to 5 (very high):

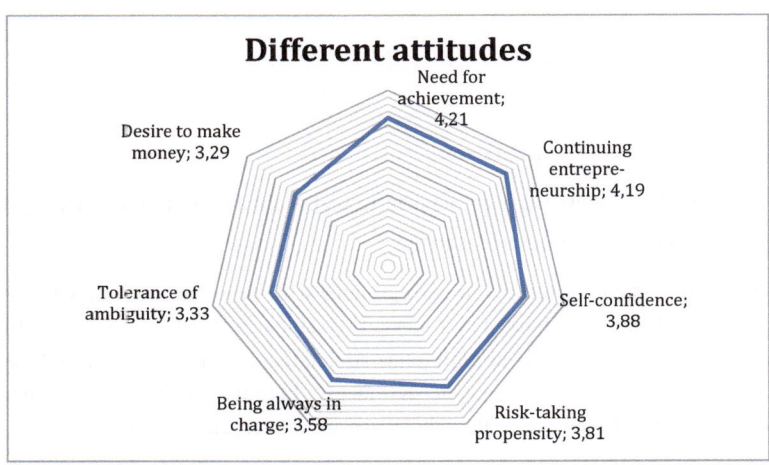

Figure 20: Different Attitudes of Entrepreneurs (n=48 (entrepreneurs))

4.7.6   Deriving Influence on the Startup's Growth

In addition to the three combinations of the different aspects of

- the average revenues of the companies,
- the number of current employees and
- the salary of the founder,

the growth of each startup is also influenced, according to the survey, by other aspects:

1) Level of personal investments
2) Experience of the founders
3) IP Protection
4) Expertise on the target market
5) Different nationalities of the team members

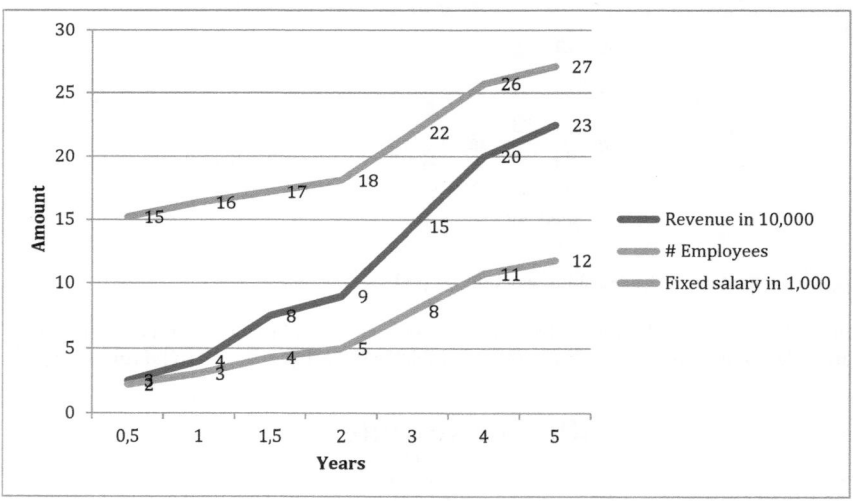

Figure 21: Average Growth of a Successful Startup

Interestingly enough, IP rights for example lead to higher salaries of the founders and more employees among the questioned startups.

Companies that had been started by an entrepreneur with prior business experience turned out to have higher revenues than others. The same can be said for teams with experiences with the target market or with international team members. The latter, the international team, is as interesting success factor, because most entrepreneurs (question 9 of Appendix B) have described their team as not very international.

The general results and rankings from both perspectives (investor's and founder's 4.6.5) are very much comparable with each other. There are few differences in the answers or rankings and most of the time the clustered answers are identical. In the following section the most important success areas will be summarized from both perspectives/ groups. According to the survey as well as existing studies, these are mainly the market, the product and the team (Spinelli & Adams 2012).

4.7.7   Most Important Success Areas

I. What are the most important market-related success factors?
The perfect market *(Both perspective)*
- Market with a high growth potential
- No big competitors on the market
- Customers who share feedback

II. What are the most important product/ service-related success factors?
Characteristics of a successful product/ service are *(Both perspectives)*
- Fulfills market's needs
- Competitive advantage
- Innovation advantage
- Quality leadership
- Protected IP (not necessary but recommended)

The perfect strategy is *(Both perspectives)*

- Quality leadership
- Innovation advantage

III. What are the most important team-related success factors?
Aspect/ characteristics of the co-founders *(Founders' perspective)*
- Previous entrepreneurial experience
- Willingness to invest their own time and money
- Good network of contacts
- Clear vision

Aspect/ characteristics of the perfect team *(Both perspectives)*
- Techy but not exclusively
- Listens and understands his customers
- Shares the vision with the founder
- Capable of reacting well to altering circumstances
- Ready to work hard and sustain intense effort
- Presents the business idea well
- International background

Aspect/ characteristics of the individual entrepreneur *(Founders' perspective)*

- Self motivation
- Creativity
- Ability to learn
- Need for achievement
- Continuing entrepreneurship
- Self-confidence
- Risk-taking propensity

The market success factors rather describe emerging markets like the BRICS (S stands for South Africa) or new areas in mature markets like the Internet has created. It is interesting to see that US startups especially in IT, Healthcare or Cleantech nowadays perform very well in America although there is fierce competition and at first glance no advantage towards Europe and Germany in terms of size, growth and maturity. German startups are not leading in these fields and have huge problems becoming global players, in contrast to the German industry domains of Automotive, Chemicals or Machine Building. The proximity to countries with growth potential is much more given in Europe and the EU itself offers a market, bigger than the US itself and still German startups seem not to take advantage of these market aspects. When examining the second success category product/ services aspects like quality, innovation IP rights or costs are essential. These are elements that suppose to be equal in the developed regions of North America and Europe. When analyzing the third success group of the team – by most investors mentioned to be the most important success factor, mainly character traits are mentioned that are pre-given and cannot be learned. A first conclusion could therefore be that the US simply has better entrepreneurs and improving an ecosystem would therefore mean to make its entrepreneurs better. Following that path, one has to identify then, which pre-given character traits of an entrepreneur can be changed or improved and what further trained skills are necessary be a successful entrepreneur (compare abilities and skills in chapter 6.4). It definitely shows however that the entrepreneur needs to be considered being a part of the Entrepreneurial Ecosystem, which is not the case in Isenberg's or the World Economic Forum's ecosystem models. And it needs to be analyzed in depth in comparison to the Booz and World Economic Forums' ecosystem. In addition, it is necessary to conduct a comparison between the German and the US market, in order to derive the differences in the market and product success area as well as to find further success factors, that explain the outstanding role model position of the US in global entrepreneurship.

5 Market Comparison

Many Americans believe that America is the greatest country in the world. And certainly America's success in business is one part of that attitude, with entrepreneurs being part of it. It is the land of business superlatives from having the most billionaires; the most venture money, the best business & engineering schools to most successful entrepreneurial hub – Silicon Valley (Jones 2005). Many nations seem to strive after the US model, and in entrepreneurship and startups talks around the globe America is still considered the holy land for entrepreneurship. A definition of a holistic Entrepreneurial Ecosystem therefore needs to observe the ecosystem of the US and probably take many of its elements into account. If there was already a holistic definition of such an ecosystem, the analysis and comparison would be much easier. This way, comparing elements have to be gradually identified and ideally lead to further relevant elements. The research and survey results however have already revealed many aspects to start with.

But whether believing that America has better and more entrepreneurs or that there must be other reasons to explain America's dominant entrepreneurial status, it still leads to the necessity of analyzing the differences between the US and Germany. Quantitative and qualitative aspects supposed to be compared, ranging from risk capital to history, in order to find the reasons for the difference in performance.

A huge difference indeed can be seen between the handling and size of risk capital. And risk capital is a good entry point for a comparison, because it touches many areas from the entrepreneur to the investor. But when comparing risk capital in Germany and the US one has to keep also in mind that most of this money is private money, often not providing sufficient transparency. Whereas venture funds provide some information to promote themselves, the business angel and family & friends investments are running very much in the background or under the radar. Only over their membership in associations and clubs it is possible to get some aggregated information on the whole market, its size and habits. That is why it is complicated to get adequate data on the number of business angels, the amount of invested venture capital or their ROIs (Backes-Gellner et al. 2012, p.90).

The different reports from organizations like the German Business Angel Network (BAND), the European Private Equity and Venture Capital Association (EVCA) or the National Venture Capital Association of America (NVCA) are not aligned and thus confusing, mainly due to contradicting data and different focus areas as well as definitions. Nevertheless, the following sections will try to show, compare and interpret the available information.

5.1 Venture and Startup Relevant Figures

Figure 17 shows that around 26bn USD were invested on average in the United States during 2008 and 2012 in comparison to a 0.66bn USD in Germany during that timeframe (also compare data from Table 8 as well as (Frazier et al. 2011, p.28)(Bundesverband Deutscher Kapitalbeteiligungsgesellschaften (BVK) 2011b, p.13)). This reveals America's venture capital business in those years being 40 times bigger than Germany's, although its gross domestic product (GDP) was only 4,4 times higher[6], with a GDP per capita in America being just 1.16 times bigger than Germany's. Eventually, the US invests 0.15 % of their GDP in risk capital and thus in startups, Germany only 0.03%. These differences are often talked about in the German startup community, however seldom with accurate numbers and reasons behind it.

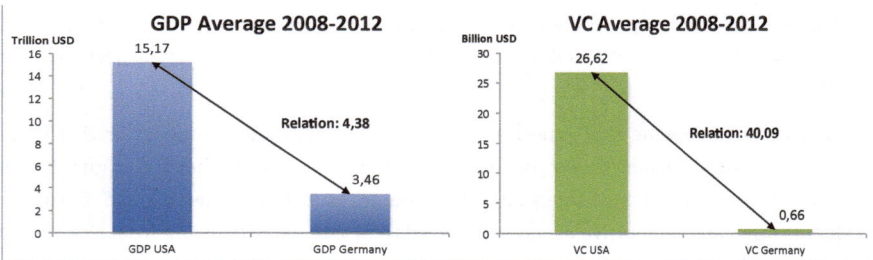

Figure 22: GDP vs. VC 2008-2012 Average (compare Table 7 & (Countryeconomy 2014)

The number of big VC funds and business angels in America also exceed the German ones by far. Germany accounts for 4 major private independent venture funds, with a volume of more than 100m USD, whereas 227 are based in the United States (H. Brandis & Whitmire 2011, p.35) – a multiple of over 56. Furthermore there are even 86 times more business angels in the US with 258,200 compared to 5,000-10,000 in Germany (compare Table 6 of section Business Angels (BA)). Given those numbers, it indicates that the average invested sum in the US in smaller than the ones in Germany. US funds invest on the one hand the largest amounts into single startups, but in the meantime they have also thousands of investors that put up only small amounts of money. This is an ideal situation for startups, because there are many investors helping to start a business in the first place with small financing and later on there are enough funds that can carry the successful startups much further with very high investments (compare Series A,B,C financing). When it comes to ROI – or IRR (internal rate of return) which most investors use a similar term – Germany in contrast is ahead, with a higher median

---

[6] In 2010 the U.S accorded for a GDP of $14.58 trillion in contrast to the German GDP of $3.31 trillion (World Bank 2011)

multiple on invested venture capital. American investors usually get 4.5 times their investments back, while German investors in 2010 even received 7.2 times (H. Brandis & Whitmire 2011, p.5). Germany is the country in Europe with the least invested risk capital of the big nations but with the highest returns. A general goal of VC funds is to return 5-10 times the invested sum of one startup investment, in order to achieve the minimum of 6-8% annual interest for the complete fund volume, with some startups failing completely or performing much lower than hoped (Spinelli & Adams 2012, p.403). Big VC funds like Accel Partners often even have the requirement for each startup they invest in to potentially pay back the complete fund volume with one exit, aiming for returning factors of 25-50 and even more (Romans 2013, p.65). A rule of thump is also, to double the fund volume every ten years, which requires an approximate annual interest rate of 8% (Manger 2011; Überla 2011).

In order to understand, where this huge mismatch between GDP and venture capital comes from and to better compare ROI performance versus total size, further data has to be compared and analyzed, which can be seen in the table below for the year 2010:

Key Figures 2010		Relation US/Ger.	US	Germany
GDP (Trillion USD)		4.41	14.58	3.31
Population (Million)		3.81	311.48	81.75
GDP/capita (Thousand USD)		1.16	46.82	40.48
Area (Thousand km^2)		27.52	9,826.68	357.11
Pop/km^2		0.14	31.70	228.93
# of VC companies		16.35	1,799	110
# of VC deals		3.47	3,294	948
# of deals/company		0.21	1.83	8.62
VC amount invested in Million USD		23.00	21.97	0.96
Median Multiple on VC		0.63	4.50	7.20
Percentage VC/GDP		5.22	0.00	0.00
# VC funds (≥ 100m USD)		56.75	227.00	4.00
Years since start of VC industry		4.00	60.00	15.00
Seed (Start-Up) (Million USD)		26.61	1,700.20	63.89
	.../total VC	1.16	0.08	0.07
Early (Million USD)		27.25	13,904.00	510.31
	.../total VC	1.18	0.63	0.53
Expansion (Million USD)		16.71	6,370.60	381.18
	.../total VC	0.73	0.29	0.40
# of BA networks		8.29	340.00	41.00

# of BAs	86.07	258,200	3,000
# of BAs/network	10.38	759.41	73.17
TEA – Total Early-Stage Entrepreneurship Activity	2.26	12%	5,3%
Self-employment (Million)	8.79	9.8	1.11
New business openings (Million)	11.49	4,71*	0,41

Table 22: Venture Capital and Economic Key Figures

Sources: (Countryeconomy 2014) (Hendrik Brandis & Whitmire 2011; Bundesverband Deutscher Kapitalbeteiligungsgesellschaften (BVK) 2011b; Förderland 2008; Frazier et al. 2011; Organisation of Economic Cooperation and Development (OECD) 2011; Prof. Dr. Backes-Gellner et al. 2012; World Bank 2011)(Global Entrepreneurship Monitor 2011)(de.statista 2014)(Hipple 2010)

* 2008-2009 10-12% being companies that have at least one employee being 0,518 million (U.S. Small Business Administration 2012)(Institut für Mittelstandsforschung 2012)

The current better ROI performance of German funds cannot compensate by far the under-financing in total, because it only serves the investors of the funds and does not provide startups with more risk capital. However, the excellent recent performance of VC money in Germany is a perfect ground for attracting more people to invest their money in startups and in VC funds. The rise of crowd financing in Germany to 20m USD in 2013 is one positive signal towards that direction (Crowdfund Insider 2014), the next generation of VC funds of existing VC companies the other. The global increase of crowd funding from 0,53bn USD in 2009 to 2,8bn USD in 2012 (Romans 2013, p.31) and the almost doubling to more than 5bn USD in 2013 however shows, that Germany with its 20m USD is again very small in comparison to the US, which accounts for 72% or 3.67bn USD (Forbes 2013b). It will be interesting to see, whether actually also more VC funds will be founded and are able to collect substantial amounts of money in the following years using the track record of their peer group. Nevertheless, with more money and funds competition among investors the ROIs will decrease due to higher evaluations, broader and earlier investment patterns. More detailed thoughts on this aspect will follow under the section of buyer's and seller's markets.

Some studies and reports from 2014 indicate also a continuous increase of business angel activity in Germany, especially in Berlin, however the next stage and growth financing is still underperforming (Die ZEIT 2014). Berlin thereby is still far behind London for example, which is the dominant city in Europe. When it comes to building up global players, the growth financing is essential and thus limits the German startup scene. Although cities like Berlin attract also international founders e.g. from Israel (Spiegelonline 2014a) the question remains how to deal with the under-financing in the next stage. Startups either need to get quickly next round financing or have to make instantly money with their business model. Since many Internet, IT and of course also high-tech companies need to reach a critical level of user bases or technical sophistication, they are doomed without ongoing financing. Many of the success cases the author knows from Berlin, Munich, Karlsruhe or Cologne have all got in contact with international investors in order to grow their company. It took them a lot of effort and time to engage a US investor on the one hand or involve a German investor with a ticket size of more than a million USD. This finding of and matching with an investor should be improved giving the startups more time for their actual business development. The under-financing of German venture capital scene is enormous and is only good from the investor's perspective. The following section will describe the market situation and explain the phenomenon of an investor's market in relation to a buyer's and seller's markets.

5.2 Market Situation

In general the US economy helps startups to grow faster and become an international player much easier, because of the size of its own economy and the role model and leading global economic position. The domestic market itself is big enough to bring a startup to a 100m+ evaluation, which attracts many investors as described in various chapters of this thesis. The international growth potential is more feasible for US startups making the investment deal even more interesting for the investors. The familiar openness for startups, failure and risk financing in the US culture not only gives US founders a head start but also continues to fuel the US Entrepreneurial Ecosystem (compare chapter 0 'The Entrepreneurial Ecosystem' in this thesis). All players are actively involved and eager to be successful, creating a positive competitive environment in the US.

Investors in Germany on the other hand can do much more "cherry picking" than their US counterparts, leaving many startups with high potential and high risk without financing. Therefore, they also perform most of the times better than their US counterparts in terms of ROI. Yet this picking is bad for the startup scene. Investors in Germany have less competition and can thus later invest in companies, waiting till those startups have shown more proof points. This is only possible in a situation, in which demand for venture capital is higher than the supply. A classic situation in business markets, leading to higher prices or in this case to higher ROIs. Therefore we can speak in Germany of an investor's market – analog to a buyer's market, whereas the US has more of a founder's market – analog to a seller's market. With the current risk aversion in Germany, a sort of envy culture and the understanding that failure is a sign for incompetence, this investor's market is being catalyzed.

5.2.1   Concept of Buyer's and Seller's Markets

The buyer's and seller's markets are two controversial market situations. Depending on the case, it is either the buyer or the seller, who has the power to determine the market conditions like pricing, payment or contracts (Cannon & Perreault 1999). The reason for the development of these situations is displayed in the following table.

Buyer's market	Reason	Seller's market
Demand < Supply	**Relation**	Supply < Demand
Non-urgent (e.g. luxury goods)	**Needs**	Urgent (e.g. accident assistance)
...on Buyer's side (own actual	**Special knowledge**	...on Seller's side (e.g. lawyers)
Seller is dependent on buyer	**Dependence**	Buyer is dependent on Seller (e.g. senior care)
No competition between buyers (e.g. luxury goods)	**Competition**	No competition between Sellers (e.g. monopoly)
Low inflation rate	**Value development**	High inflation rate

Table 23: Reasons for Strong Positions in Markets (Wirtschaftspedia 2012)

The concept of buyer's and seller's markets can be used and adapted for almost any free market e.g. real estate or consumer goods. Through the financial crisis in 2007/2008, the overheated real estate prices in the US dramatically fell (The Economist 2013). People were forced to sell their homes, because they could not pay their mortgages and debts back, thus swamping the markets with houses for sale and increasing the supply for houses. On the other hand, demand was going down, because people did not get good credit rates for buying houses, they lost or were uncertain about their jobs and they were afraid to buy a house in a financial downturn. With all the houses on the market, also real estate investors were holding back, because of falling prices. All this turned the market from a seller's market, where everyone wanted to buy houses, to a buyer's market, where the few people that were buying houses, were getting huge discounts. Interestingly, this movement also hit the high value houses and not only the standard family homes.

The concept of buyer's and seller's markets can also be transformed to the venture capital market. The most important thing to define is, who the buyer and who the seller is. Since the VC is investing the money in exchange for equity of a startup, one can say the investor is buying the equity, thus becoming the 'buyer'. And because the entrepreneurial team is offering shares of its company and eventually selling them, they become the 'seller'. Similar to the financial crisis and the breakdown of the housing market in the US, the burst of the Internet bubble in the year 2000 had a similar effect on the startup market. In the late 1990s Internet startups in the US and in parts also in Germany were flooded with money, without the need of a track record or a stable and proven business model (Beattie n.d.). It was a perfect seller's market for the young ventures in terms of receiving finance. Investors were afraid of losing a potential blockbuster deal and thus were deliberately given away the fund's money. At that time the venture capital volume in Germany reached its highest point of 4bn Euros in the year 2000 (Frank 2006). After the burst the volume was falling to its lowest point in 2003 of only 0,5bn Euros. It turned the market everywhere into a buyer's or investor's market, what it unfortunately still is in Germany. However, the situation in Germany and Europe was getting better, although the financial crisis in 2007/2008 has slowed down and cut of some of this improvement (H. Brandis & Whitmire 2011, p.14).

As in other economic situations, the US market is always faster with its reactions and recovery; it almost stayed the whole time a startup's or seller's market, with short time exceptions (CNN 2013).

5.2.2   Situation in Germany and the US

Egbert Freiherr von Cramm (Managing Director of Deutsche Bank) describes the situation in the European venture market as a buyer's market (Hendrik Brandis & Whitmire 2011, p.18). He further mentions that the "[...] *European venture capital is a cottage industry with insufficient number of private investors* [...]". He believes this is due to lacking pension and endowment funds, which make up 65% of the US venture capital.

Germany is not Europe, however it is possible to take the German market as a good representative of the European market, because of its size and influence in Europe. Since the returns and the risk aversion are even higher than those of the other European countries, von Cramm's thoughts fit even better for Germany. Two main issues are thereby the case for Germany:

- The capital supply for startups is much lower than their demand (H. Brandis & Whitmire 2011, p.12)
- Caution and risk aversion about the development of startups lead to later stage investments

Those two points stand in contrast to the American situation, where capital supply is almost at any time higher than the demand of the startups, and where American investors are more open to bear risks very early.

5.2.3   Results of the Different Distribution of Power

The VC competition in the US has a very positive influence on the development of new technology, job creation and economic prosperity. In order to fulfill the promises to the owners of the funds to double their money within approximately 10 years, the VCs always have to look for many new investment opportunities and thus invest in many different and also more risky ventures, because there fund volumes are so huge and need to get invested. The amount of deals and the ticket size are therefore much bigger than in Germany. On the one hand this also leads to a finance funnel at every stage of a company including the very Early Seed and Pre-Seed phases. On the other hand it also extends the spectrum of industries that receive funding. Besides the wider and probably looser investment habit, the competition on the other hand professionalizes the investors to be able to provide the startups with profound industry knowhow and highly motivates them to bring in the contacts and experience. All in all, these high volumes of investments increase the chances, of financing a blockbuster technology and enterprise. That is probably the reason, why the American industry so far has always been able to be at the edge of the latest trends and hosts the most modern and biggest companies.

Additionally, looking at the figures, a little lower return rate applied on a way higher amount of invested money still makes a higher absolute amount of gained returns. Therefore, more German investors supposed to be attracted by the high multiple, which is currently being able to be achieved and which is higher than the one in the US. Uli Fricke – Chairwoman of Earlybird from 2010 to 2011 – explains that the insufficiency of venture capital in the EU and therefore in Germany has on the one hand led to low entry valuations, but on the other hand driven up the capital efficiency, which suppose to be about 70 percent higher than in the US, due to the fact that the few investors can simply be more selective" (Hendrik Brandis & Whitmire 2011, p.19).:

Till the point is reached where too many investors in Germany increase evaluations and push themselves for earlier and more risky financing, there will be high ROIs also for new investors and funds for quite a while.

Also, as investors in Germany have the power of price determination they get relatively big shares for small amounts of invested money. According to some unofficial VC sources, this leads to smaller entrepreneurial activities because entrepreneurs get discouraged after their first conversations with investors and the huge drain in equity they would have on their own companies, when a German investors starts to invest. This effect was increased after the Internet bubble busted in 2000, when the evaluation given by investors had dropped almost 3 times (Hendrik Brandis & Whitmire 2011, p.37).

Last but not least, as the number of potential investors is so low, there is hardly competition among them, thus the question, whether a startup will receive fresh capital, strongly depends on one subjective meaning. In contrary, the American venture capital market with its huge competition generates more investments automatically because of the peer pressure. Assuming that a VC company rejects to support a startup team, it always has to be afraid that a competitor makes the big deal with the billion-dollar idea and thus the VC has lost a chance to fulfill the promise made to the fund's investors for a high ROI. This type of peer pressure does not exist in Germany, because of the low number of VC funds. Therefore it happens quite often that startups in the US can even choose between different VC firms for the best contracts, whereas in Germany many teams will not receive funding at all.

5.3   Possible Reasons and Analysis of VC Developments

When comparing the American VC market at a time like 2010 in which US investment was not as its peak with Germany, it was still 23 times bigger than the total VC investment in Germany. Multiplying the factor 4.4 of the higher US GDP with the 4 times longer history of venture capital in the United States – assuming that within those approximately 40 years of earlier VC the investors have stayed within the investment system and within the habit of a 10-year investment cycle and thus having at least 4 generations of VC funds – one already obtains a multiple of 17.6. Unfortunately this numerical approach has its boundaries for the pure multiplying works only for stochastically independent factors. Obviously, the given factors are not independent, because, for example, successful investors automatically attract more investors or an increasing GDP may also increase the amount of wealthy individuals, and thus increase amount of total investors. In addition, having more investments for startups leads to more startups and normally to more successful companies, whose entrepreneurs can become investors themselves. The VC system therefore can feed itself and grows over-proportionally. This could explain the factor increase of at least 23 or even higher than 17.6. With the higher amount of success cases and general economic wealth increase, the US has created a more than linear growth, with a growth factor higher than 4 in in 40-years longer start of VC history than Germany. New investor groups like pension funds are also an explanation for the additional increase. The numerical approach by itself cannot explain or justify the big differences in the venture capital market of both countries.

In this context, non-numerical factors, such as for example the entrepreneur's' reputation, risk aversion or VC experience, play a role. They are considered to influence investments, but cannot really be measured in numbers. Therefore, further qualitative aspects have been picked, in order to explain the differences in the VC activity.

5.4 History, Geography & Culture

In order to understand the difference between America and Germany better, one has to compare the influencing factors that have formed both societies. This requires an analysis of historical events.

Many early and later American immigrants were persons that left their country for religious, political or economic reasons. In the early 17th and 18th century, leaving a country and crossing the ocean in search for a new, free beginning carried great dangers. To those individuals and their families, taking great risks at every step of the way became a way of life. It formed the American attitude, spirit and character then and still today – especially in business. The entrepreneur is product of this unique frontier culture, which can be found in America, as well as many Spanish, French and other English speaking countries, where Europeans migrated to since 1492. This typical American attitude and stereotype is the base America's entrepreneurial culture today (International Monetary Fund (IMF) & Mihet, Roxana 2012, pp.4, 36). A gene pool of many people developed, with a tendency to openness the willingness to change or try something new, even if risks were involved. The overwhelming size of America, the lacking infrastructure and dangers forced many emigrants to work together, while at the same time they had to take their fate in their own hands. In its early years, the US also offered thousands of opportunities to discover, and gave people the chance to implement new ideas and projects. For these and other endeavors, a never-ending stream of people from different countries and cultures immigrated to the US, and thus were a source of inspiration and implementation. With a rising success story, it attracted even more foreigners to come to the United States and to participate in the general overall country and specific business development. Parts of this can be observed in the American Dream. From the beginning up until today, it is the mixture of internationals that drive the American industry and business. In 2014, 40% of America's Fortune 500 companies have CEOs or founders, who are not native US Americans (Forbes 2014a). The innovative hubs around the universities of Stanford, MIT or the Universities of California are filled with students and persons from around the world. After the total World War II destruction of everything in Germany and the brain drain to America, the country had to start all over again. Tens of thousands of entrepreneurs, with their backs against the ruins, had no choice but to take the risk and chance to build up something new.

5.4.1   Anglo Saxon Tradition

From colonial times up to the present heritage and mind patterns, the Americans were and are strongly influenced through Anglo-Saxon language, law, business, philosophy and culture. Especially many of the early US leaders were educated and influenced by a British tradition. Although many of these first Americans from the European continent were trying to free themselves from British or European laws and regulations, they did not change their habits and culture. The philosopher John Locke (1690) and the empiricism of the 17th & 18Th century in Britain moved away from the classical, absolute

European thinking of ideas and being that started with Plato and Aristotle (Hersch 1989, pp.140–143). Locke's understanding of the human perception of things led to a very pragmatic attitude towards life. For him, reality exists only of the so-called primary qualities (or properties) like firmness or movement, independent from human perception. Since ideas are part of the human mind and not based on reality, they are not part of the "primary qualities" of philosophy. They originate through personal experience, thus all perception and experience depends on the individual's perception of reality (Locke 1690). For the philosopher and the realists, absolute thinking and laws in society are, therefore, against movement or change of things and potentially even exclude vital aspects (Hersch 1989, p.141). This thought structure of embracing reality through experience and not through pre-given ideas and absolute being has its influence upon modern case law, liberal market theories and political thoughts in Britain and the US (Mace 1979). These historical and philosophical aspects have had great influence on American society, politics and business – especially upon entrepreneurship and risk capital in the US.

5.4.2 Geographical Position of the US

In addition, America's geographical position has been nurturing and developing these effects. Once the territory had been completely conquered and the borders had been fixed, the USA became the "biggest island in the world". Two oceans separate it from the rest of the world, leaving only two borders to neighboring countries in the north and south. The one to Canada is rather a "soft border", with people very much alike, while only the border in the south, to Mexico, stands for a gate to a different world. Within the US, people can fly or drive for hours without getting in contact with Non-Americans. This rather isolated geographical situation is also influencing the American culture very much, yet it does not seem to have a negative influence on entrepreneurship.

Despite varying aspects within the 50 individual states, business can be carried on quite easily, due to the huge homogenous market reaching out to more than 300m customers from the East- to the West Coast – with one official language, the common world currency, a systematic developing and expanding business attitude. Thanks to the American Federal Government system and the numerous federal reforms, as well as the integration of millions of foreigners, carrying on business especially for an entrepreneur or startup is much easier. Even today, the Mexican and Hispanic Community, which still has, at least in part, integration problems, nevertheless creates an additional US market of 1.5 trillion USD (Kelly 2014). Low paid agricultural jobs, as well as a buying customer base of telecommunications, clothing or entertainment contribute to this additional market. Regardless of whether you are an immigrant or a Native American, many believe in the American Dream and the Pursuit of Happiness for each individual. These aspects and considerations led to the typical American self-understanding that everyone in the world, outside of the US, has become aware of due to the newspaper reports, movie and television entertainment, as well as book and journal publications. This also explains, why Americans have such a strong sense of pride in their country and in their countrymen's success in fulfilling the American Dream.

The geographically isolated position of the United States in combination with the liberal, free market policies and the general dangers throughout the urbanization of the North American continent from the 17th to 20th century forced individual and collective adaption, as well as to help themselves while also helping each other. Thus, over the centuries a certain self-understanding and focus on the American territory and market occurred. Due to the emphasis on a common credo and traits, as well as its strong patriotism, unification, through federalism, in business and other areas is more developed and advanced than in Europe.

5.4.3   Geographical, Political and Cultural Situation of Germany and Europe

Up until the mid 1950, due to wars, international job migration, changing cultural and religious factors, as well as the significantly diversified rather than the unified economic and cultural society, there was no European attitude and thinking. The heterogeneous background and insistence of state sovereignty created many more hurdles for unifying the European continent – than the North American. The greatest burden can also be the greatest chance the same time for business, regardless on which continent it occurs.

A common European spirit, unfortunately, is slowly growing, due to the language, cultural and legal barriers, which make business operations in a single, homogenous European market more difficult. Eventually, the political attitude of the American economic elite that the government should not have too much power and influence on the market, in combination with the strong reluctance of socialism and communism, encouraged US business society to help itself rather than wait for the government to do so. The fact that some European court decisions or laws are very difficult to change, due to the diversity of national interest and fears of their lobby groups – for example French farmers protesting against changes of European subsidies – gives American entrepreneurs clear advantage, although US laws are also hard to change, especially where the interest of lobby groups are affected. Nevertheless, in the US there is one national interest, instead of many diverse national interests in Europe. In addition, US businesses try in most cases to lobby and fight for retaining and expanding free markets, which is generally positive for entrepreneurship.

5.4.4   Current European Innovation Initiative

Nowadays, Germans and Europeans might look for EU funds to get entrepreneurial activity started, instead of starting out by themselves. The EU, under the former president of the EU Commission José Manuel Barroso, in 2008 initiated a 300m € program for more innovation, through creating a European Institute of Innovation & Technology (EIT), with a 2.7bn program until 2020 (European Institute of Innovation & Technology 2014). This is an important financial effort by the EU, however, there is a lack of initiative within the entrepreneurial community, because the politicians and academia did not consider the main actors – the inventors, the entrepreneurs and the investors – adequately, when they created this institution. A very important ingredient for entrepreneurship and risk financing is not to depend on a public net to secure one's ventures and then wait for the approval by the government. This short come undermines at least parts of its nature. Government programs should encourage the desire of the main actors of entrepreneurship and not lead them. The pattern of fostering entrepreneurship and innovation a great extent in Germany to comes from a theoretical perspective. In the US on the other hand, entrepreneurship hubs or centers were mainly created by entrepreneurs, who collected the necessary funds from private individuals – mainly alumni. The German concepts are often initiated and steered by university officials that collected money from several government programs. The entrepreneur was too often facing and discussing the business idea and development with persons, who have never founded a company in their life. Although these persons are very often motivated and mean well, they lack the experience and mindset from the entrepreneur's side. Nowadays, many entrepreneurial programs try to integrate industry experts in their panels, for the design of new concept, as well as for the implementation phase, in order to be more professional. However, they still have to prove themselves.

5.4.5   The German "Kleinstaaterei"

From a historical perspective, it must be remembered that for over 1,000 years Germany in contrast to the US, being in the middle of Europe, was exposed to numerous traditions, wars, various languages and interests. In the 17th century the "Heilige Römische Reich Deutscher Nationen" still consisted of about 300 independent states (Kulke 2011), while up to 1806, until Napoleon dissolved the 1st Reich, it consisted of *Stadtstaaten* 'City states'. This coined the German term *Kleinstaaterei,* standing for many very small states having different laws, regulations, and opinions, although their proximity as well as language and cultural similarities should rather stand for unity. About 60 years after the "new" founding of Germany in 1949 (Deutscher Bundestag 2013) and 20 years after the German "Wiedervereinigung", it still consists of 16 states, although in total Germany with its 357k km^2 is smaller than the state of California having 424k km^2 (Wikipedia 2014a; Wikipedia 2014b). And even within these 16 states today, there are still many differences, due to the historical backgrounds and different traditions as well as beliefs. One could say the homogeneous group of Germans has inherited to think as a heterogeneous one. Every small or large state, region or city needs to have their own political representative and different opinion. Thus, from a German perspective, it is therefore often difficult to speak with one unified voice; on the European scale it is next to impossible, because the other European nations also have a similar fragmented domestic situation and history. This is the main obstacle for Europe to grow stronger together and form a similar unified market as the United States of America.

5.4.6 Honor and Pride

With Germany being in the center of Europe, it has been geopolitically speaking a transit country and often was in the middle or involved in almost every war on the continent. Through the changing rulers and occupants, a certain cautiousness, social cohesion to a close group or village and an envy culture might have been bred, ignited by destruction and hate against the suppressors or the persons that were profiting from it. This stands in contrast to the American attitude, where its citizens generally were proud of the success of others. Professor Miller *(1993, p.86)* from the University of Michigan supports the author's line of thought when he writes in his book Humiliation that *"Honor is above all the keen sensitivity manifested by the desire to be envied by others and the propensity to envy the successes of others.".* The foreign, victorious occupants had been tackling and disgracing the honor of many German inhabitants, and thus nurtured the German envy characteristics over centuries. Of course this topic is certainly much more complex and many further aspects have led to stereotyping Germans of having an envy culture.

Alfred Adler (1996, p.196) one of the founders of the *Individual Psychology,* connects the envy trait to inferiority complex as well as factors like suffering, deficits and social problems. He also explains how envy is depending on the relation between individuals and groups, where the better position and power of one side leads to a reaction and freeing of forces on the other side. Character traits usually are developed during the childhood, which can be an explanation, why this envy trait seemed to be inherited by generations of Germans. Once the parents had been unhappy, suppressed and envied others, they transferred this trait to their children (Whalley 2014). Given the situation that there were constant sources of war and suppression, there was always a base for a new envy development within the next generation.

5.4.7 Educational Systems

Another hindering factor for an entrepreneurship culture, could have been the German education and social reform initiated by Bismarck (Albisetti 1983). The idea of disciplined and well educated Germans combined with the concept of a welfare state, does not really aim at building an elite and fostering the individual strengths of its inhabitants. Entrepreneurs in contrast are living from the fact that they want to make the best out of their own strength and become an elite; an elite that also leads people and creates jobs. This type of educational system can rather be found in the US, where the building of an elite is intended. Through the early focus in high school, the US tries to strongly promote talented people, whether it is in science, sports or art. This way, strengths are further nurtured, while the idea of a broad education for all students, like in Germany is partly abandoned, till the students enter their first college year, with a broader educational focus (Study in the USA 2012).

5.4.8 Beginning of Entrepreneurship and VC in Germany

Despite the national difficulties, the beginning of entrepreneurship and industrialization had been in Europe. Aside from Britain, Germany brought up many big companies and became a successful business nation in the 19th and early 20th century (Pierenkemper & Tilly 2004). Even with an envy culture and more socialist emphasis, Germany became an entrepreneurial nation during those times. After World War II Germans started from scratch and thousands of entrepreneurs took their chances. It seems like entrepreneurship also comes in waves and influences society in its time. Germany at the moment seems to be in the lower part of entrepreneurship wave, rather focusing on holding on to its wealth and business structure rather than reinventing and renewing itself. Maybe a rise of entrepreneurship in Germany will soon become greater.

When analyzing the venture capital industry in Germany, however, there was too little money after the war – only debts and destruction all around. Not the best circumstances to start the VC industry, because VC structures always require resources with an abundance of money, e.g. coming from wealthy individuals, thriving companies and partly from the government. These types of investors in Germany and mostly in Europe were busted due to World War II – some already after the end of World War I. It was not before 1975 that the "Wagnisfinanzierungsgesellschaft" ('Risk financing corporation') was founded as the first German VC-company, in which mostly credit institutes participated. A next milestone was the first VC fund, initiated in 1982/83 with a maximal height of 20m EUR. In the 80s and 90s VC companies like "Technologie Holding" or "TVM Capital" came to be the leading representatives on the private equity market (Jacobi 2010).

5.4.9   The Beginning of Venture Capital in the US

The mentioned factors in this chapter and especially the post war consequences for each country, gave the US a decisive advantage to install VC structures much earlier than Europeans. When VC started in the US after World War II (Jacobi 2010), Germany had to start its general economy from zero, concentrating on classic rebuilding and basic entrepreneurship. The US, in contrast, was able to build and try out new financial models upon its existing strong economy.

Jacobi (2010) further explains that in the USA the financing form *Development-Capital*, as the VC were referred to World War II, already had emerged in the first half of the 20th Century. Mostly wealthy private persons or families in the first generation invested in young companies. For example Laurance S. Rockefeller in 1938 helped to finance Eastern Air Lines and Douglas Aircraft. Before and after the Nazi regime, these early forms of risk financing did not occur in Germany.

The first two VC companies founded in the USA right after the Second World War were called American Research and Development Corporation (ARDC) and J.H. Whithey & Company (Romans 2013). Thanks to their long history, some VC-companies, today also have a strong tradition and plenty of experience in the field of entrepreneurship. For example Greylock Partners, founded in 1965, invested in several hundreds of startups and gathered experience from more than 150 IPOs (Jacobi 2010).

The earlier beginnings of VC in the US also led to the establishment of the "The Small Business Investment Act" from 1958, which boosted the VC scene throughout the US. It officially allowed the SBA (US Small Business Administration) to give licenses to private investment companies that helped financing and managing small young businesses in the US. It declares the policy of the Congress and the act itself to have the intention to improve and stimulate the US economy and in particular the small business segment through programs that encourage private equity and long-term loans funds to finance these SME and their needs for growth and modernization (US Congress 1958, p.5).

5.5   Consequences of Historical Differences

The historical differences in the two countries led to several other distinctions. Due to the young tradition of VC in Germany, only a few overwhelming success stories have been created, till today, to spread the word and stimulate the entrepreneurial scene. The existing success stories in Germany center around mainly post war companies that have become giants, but required many decades to become that large. In Germany after the war, one would often be advised to start slowly and think small. In contrast, today's successful startups, however, often need to grow very fast due to their international competition and globalization itself. The US attitude fits better in this business context advising to think global yet start small or local – a fine difference with a huge effect. This approach is further being described in chapter 7.1.4 on scenario planning (Timmons & Spinelli 2004, p.91).

This act, along with social and technology changes in the 1960s, fueled the desire of American entrepreneurs to found their own company. This act was one of the first catalysts to vastly increase this industry on both sides – the founder's and the investor's side. Another mentionable point is the reputation of entrepreneurs in the observed countries. Whereas in Germany entrepreneurs often are classified as persons, who would not have succeeded elsewhere, in the US they are popular heroes who really made it (Baker 2003, p.2) and have built a fortune out of nothing. Persons, like Bill Gates or Mark Zuckerberg, embody the traditional American Dream, which actually became an entrepreneurial dream, in which individuals became entrepreneurs, businessmen or other successful stars rather than statesmen (Baker 2003, p.3). This definitely has also a psychological influence on the VC market, when the investment is for or potentially creates a hero.

As explained, many of these factors have their origin in historical events and behavior, or they have simply started in the past. Besides character traits, *Kleinstaaterei ('Small State Pride')*, pre- and postwar developments or the educational systems there are also differences in the tax and legal system, which need to be further scrutinized. Furthermore, as part of the character traits, risk aversion also requires some further investigation, because it is one if not the most mentioned arguments where Germany and the US differ in.

5.6 Risk Aversion

"The biggest difference between the German and the American startup-culture becomes apparent through the risk disposition [...]." (Maier 2011b) The risk attitude of Germans seems to be caught between perfectionism, envy and general perception of failure. But it all originated, when the risk aversion is due to another stereotype, saying that Germans do not like to talk about their money or their earnings. This shows the lack of celebrating success on the one hand and envy on the other hand. Often different salary levels lead to resentments if talked about. It also shows the caution of Germans to talk about failure and success, because once a statement on money, salary or revenues has been made, reducing or evading it is not possible anymore. It further underlines that Germans hardly talk about their problems, failure and losses, because there are seen as a sign of weakness and even possibly stupidity and incorrect handling – the opposite of the perfectionist German world. This combination leads to low social reaction from the personal environs in case of success, even though usually humans, as social beings, want to be rewarded for their achievements by others. Especially when a high risk has been

successfully dealt with, recognition is desired. The *social return on invest* might therefore be low. This means that socially any risk activity rather brings out a negative result. Therefore, Germans remain silent about their business ventures, before they have positive results to report. And possible because of the envy culture, success will be modestly talked about. Furthermore, due to the tendency for perfection, Germans generally remain skeptical about their and other people's ideas, till the absolute proof of success. All these stereotypes together are not ideal for entrepreneurship, which needs constant interaction, feedback and motivation from others help entrepreneurs to improve and sustain in hard times. Perfectionism is also very difficult to attain for a startup, because most processes, services and products not fully developed, well tested or established in the market. Of course this skeptical environment in Germany (Maier 2011a, p.8) is unfavorably for entrepreneurs because *"the chances of getting credits or venture capital with a not fully developed business plan are small"* (Maier 2011a, p.9).

In the end, this almost leads to a fear and stigma, if one has failed with his or her business. Whereas the stereotype for the US is that an entrepreneur will mature and can learn from the failures, thus be more successful in the next venture (Iwata 2012). This fact is even supported by the "fresh start" bankruptcy law, which will be further discussed later in section 5.10. The different attitudes are also reflected by corporations and through their behavior. Trying out something new and thus risking reputation and even losses is much more accepted in the US. This is very helpful for US startups, because they often get a chance to sell their products and services to established companies, whereas their German counterparts seldom get this kind of opportunity to the sales group.

This open-minded attitude for new things and approaches is also helpful for the corporations. Companies like 3M, Google, Microsoft or Apple – which are considered to be two of the most innovative companies in the world – work with concepts known under the title 'Time Off' (Maier 2011a, p.8) or 'Blue Sky' (Digital Trends 2012). Thereby employees are allowed and even encouraged to spend part of their working time on their own ideas and concepts, which can support the company's innovative development. At Google it was up to a day per week for all employees, at Apple only selected employees receive a few weeks off to work on their own ideas or projects. At Microsoft the "Garage" gives engineers even physical space to work on other projects using Microsoft technology (FastCompany 2013). Products like Google Maps, AdSense or Gmail are said to be the results of these *Corporate Entrepreneurs*. Of course these programs are not only intended to boost innovation, but also to attract and hold the best employees. Nevertheless, Google withdrew its 20% rule of free time for any project in 2013, because it needed the time back from its employees for the running business and its new projects and products that were already in development or planned. Nevertheless, it shows how open and flexible US companies can be when it comes to new approaches, styles and concepts. This attitude helps entrepreneurship a lot. It is either directly fostered through allowing people to start and work on their own ideas, or it nurtures the climate for a pro-Entrepreneurship culture also in the corporate world. One important aspect is that companies accept the risk that their employees might create their own companies and never come back from the personal projects they are working on, or they return without concrete results for the company. This indicates another piece of the lower risk aversion open-minded attitude in the US. Such corporate initiatives are rare and rather unfamiliar to German companies.

5.7 Boldness – The Global Market Size Estimation

A combination of a lower risk aversion and more boldness in the US can be observed in the market size estimation of US startups. A very common technique is the analysis and estimation of the total available market (TAM), followed by the deduction of the non-serviceable areas in terms of price, service, technology, geography, laws etc. (Stanford University 2008, p.2) Finally, a startup should then calculate and guess, how much of the serviceable market it can conquer. While it is for US startups absolute normal to show the available market and the serviceable available market on a global perspective, in, on the other hand, a German startup it is considered to be unrealistic, if they do not focus on the German market and show primarily the national numbers. German startups experience that at business plan competitions, from business angels or from venture capitalists. The US estimation approach can easily be accessed through the Internet, where it is taught at universities like Stanford or MIT, and pitch decks of some of the best US startups can be found using this estimation style. A good example is the private home rental place Airbnb, which considers its total available market to be bigger than 1.9bn trips booked, leading to a serviceable market of 532m trips, matching their service criteria of rather low cost budgets travelling that could be done online. They consider their market share to be 10.6 million trips with an earning on average of 20 USD per trip, and thus estimate sales of 200m USD (Airbnb 2008). A calculation like this would have probably been laughed at in Germany. In 2013, Airbnb was valued more than 2.5bn USD and the most successful player in its field (Forbes 2013a).

Market Size   5

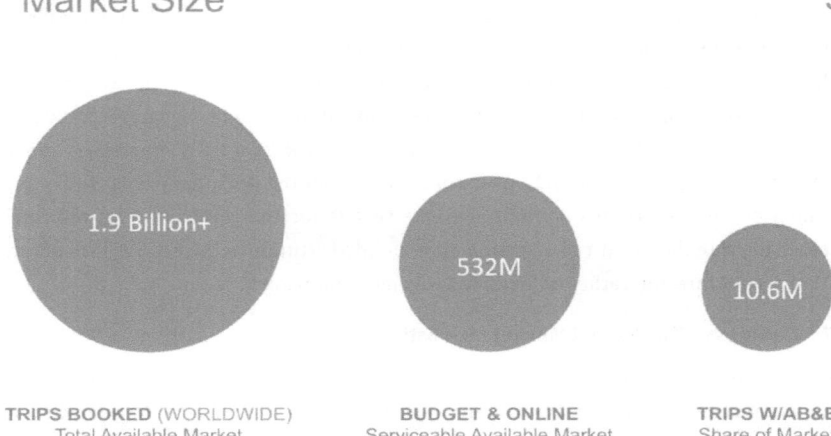

Figure 23: Airbnb Pitch Presentation (Airbnb 2008)

After the estimation or research of the market size, the startup can quickly do a "top-down" business case calculation using the market size numbers (Gabler 2014b). A thoroughly calculated "bottom-up" approach will then be used, to calculate the profound business case, with all internal expenses like personnel, office or travel, all sales numbers, a classic profit and loss statements, as well as cash flows (Gabler 2014a). Top-down and bottom-up approaches or calculations are typical for market size estimations and business cases. They are a common tool in management consulting and should be explained in detail at the university (Hoopes 2003).

5.8   Performance and Maturity of Venture Funds

The higher risk aversion in Germany resulting in few investors is still very complex and quite common. One could assume that stories and experience of many failures could be another reason for this backwardness. In spite of this reluctance and looking at the venture exits of Germany in 2010, one can see that with $4.4bn it is the leading society in Europe (with a total of $13.8bn) and has achieved that with a smaller share of investments of just $0.8B, than the UK or France (Hendrik Brandis & Whitmire 2011, p.4). This means that Germany has with 5.5 the best multiple on invested cash in Europe (European multiple of 3.5). And the median multiple on invested cash with exits over 100m USD lies in Europe at 7.2 topping the one of the United States with just 4.5 (Hendrik Brandis & Whitmire 2011, p.5). Even with all the drawbacks in Germany, it is the most efficient player in Europe, which also indicates the success of German VC. Taking these facts into consideration it is still quite incomprehensible that there is so little VC activity in Germany compared to the US. However, things can change for the better within the next years, if one takes a look at the number of VC fund generations. The long tradition of VC in America demonstrated that in the US, there are many more funds and fund managers in their fourth and fifth fund generation (Hendrik Brandis & Whitmire 2011, p.17). European funds and fund managers, due to the short history of VC only have a few in their fourth and fifth generation. But once these funds move from the second and third generation forward, there will be many more experienced fund managers and funds in the fourth and ultimately also in the fifth generation.

VC teams' maturity by number of funds raised	W. Europe	U.S.	Ratio
≥2	73	334	4.6x
≥3	58	202	3.5x
≥4	28	132	4.7x
≥5	8	94	11.8x
≥6	4	65	16.3x
Time since early growth of industry (years)	10-15	50-60	4-5x

Table 24: Number and Maturity of VC Funds (Hendrik Brandis & Whitmire 2011, p.17) Original: Cross selection of Dow Jones Venture Source

VC funds will, therefore, play a decisive role in the change of risk handling and aversion in Germany. Major obstacles, in the way of maximizing profits and earnings for investors, come certainly from taxes. In a 2010 poll implemented by Deloitte et al (2010, pp.18–20) tax policies appear to be among the most positive or negative influential parts on capital climates. According to these figures, the taxation is a bigger issue in Germany than in the US, and is therefore more likely to hinder than to foster VC and thus entrepreneurship.

% agreeing on the following theses	Germany	USA
Effective tax policies that encourage risk taking and investments create favorable VC climate	61%	21%
Unfavorable tax policies create non-favorable VC climate	72%	59%

Table 25: Tax as a Factor on VC Climate (Deloitte et al. 2010, pp.18–20)

5.9 Taxation

The numbers from Table 25 above show that taxation systems can have a big impact on the entrepreneurial and investment activity. Investing money in Germany with a ROI potential currently being higher than in the US for example, and then paying more taxes could even or lower the advantages of better investment multiples. The "Expertenkommission Forschung und Innovation" EFI ('The expert commission of research and innovation') comes to conclusion that the weakness of German VC is not coming from a lack of investment opportunities but from the struggles of the funds in collecting money and their tax limitation in their investment activities (Backes-Gellner et al. 2012, p.90).

Besides missing exit channels and top success cases, taxation could be considered one of these difficulties, and therefore requires further scrutiny in the following.

The taxation systems, in both Germany and the United States, are often said to be very confusing and chaotic (Piper 2012). Due to the high impact on an economy on the one hand and the emotional reactions among voters on the other hand, taxation issues are always topics in all political campaigns. And so regulations are constantly under close scrutiny and subject to debate or change. Not only the complexity in general is being discussed but also other aspects like the linear, progressive or proportional rate levels; advantages for business owners; taxation of profits from equity deals; estate tax and the general fairness of the system.

In order to compare the taxation models in different countries, one has to understand that due to the complexity of the system, a comprehensive view is nearly impossible. Furthermore, comparing and using pure numbers needs to be done with caution (Bundesministerium für Finanzen 2009, p.7). When for example, some conditions for a certain group or case might be advantages in one country, another tax for a following or related topic might even lead to a disadvantage.

An additional difficulty arises especially for the USA, where its federal tax system co-exists with that tax system of 50 independent states. Because an analysis of all of them would go beyond the scope of this thesis, one can only investigate selected states for the comparison with Germany. They need to have a close relation to entrepreneurship, venture capital and business development. California, as the home of the Silicon Valley and the world's leading venture region, is one, Massachusetts with MIT and Harvard, the state, which leads the US in education, is America's second most active venture region (Thomson Reuters 2014, p.14). Delaware, known for its lax 'Delaware Corporation' laws and famous as the 'state without corporate taxes' and financial home of 60% of America's most successful companies, will be taken as the third representative for this thesis' tax comparison.

5.9.1   Income Tax

The income tax "is the tax on the income of private/natural persons" (Kreft 2011a, p.1). Thereby, a certain percentage of the personal earnings have to be paid to the state. In both considered countries one counts as fully taxable as long as one lives and receives income in the respective country (Kreft 2011a, p.1) – another case is not considered in this thesis.

5.9.1.1 Concept and Comparison

The actual calculation of how much is to be paid, takes place in two major steps. First, the subject of income tax is to be determined. Hence, in both countries one deducts tax from the full income for several expenses such as education, medication, allowance and other tax exceptions (Department of Treasury (IRS) 2006, p.142ff; Kreft 2011a, p.6). Second, the tax rate depends on the subject's taxable of income, as shown in the tables below.

Income Tax rates in Germany

€-Range	$-Range	%-Range	Ascent linear...
0-8,004	0-10,556	0	/
8,005-13,469	10,557-17,763	14-23.97	Progressive
13,470-52,882	17,764-69,741	23.97-42	Progressive
52,883-250,000	69,741-329.700	42	Proportional
250,001- ∞	329,701- ∞	45	Proportional

Table 26: Income Tax Rates in Germany (Kreft 2011b, pp.185–186)

Income Tax rates in the US

$-Range	%-Range	Ascent linear...
0-8,500	10	Proportional
8,501-34,500	15	Proportional
34,501-83,600	25	Proportional
83,601-174,400	28	Proportional
174,401-379,150	33	Proportional
379,151- ∞	35	Proportional

Table 27: Income Tax Rates in the US (Department of Treasury 2011)

5.9.1.2 Differences and Potential Entrepreneurial Impact

In Germany only the federal government collects income taxes, whereas most of the American income receivers have to pay federal and state income tax, other taxes to the municipal government. According to The Department of Treasury (2011) the average state collects an additional tax, at the rate of 6.7 % while the states Alaska, Florida, Nevada, New Hampshire, South Dakota Texas, Washington and Wyoming spare income tax completely. California (7.39%), Massachusetts (6.3%) and Delaware (8.25%) compared to the states are considered the "expensive" states. These three states are at first glance a disadvantage for entrepreneurs, since they cause more complexity and higher personal taxes. However, since America's most successful new ventures and most active venture communities are in California and Massachusetts (Florida 2013), founders must for go for some other positive reasons the advantage of founding a company in a US state with lower or no state income tax. On the other hand, the German way of linear progressive rate ascents also leads to a certain complexity and generally a higher income tax, consequently there is no real advantage in the end. In comparison to Germany, the flexibility in state taxation could even be seen as an upside of the American system, leaving more freedom to the states to attract startups and founders to chose the state for their incorporation. For those startups in low or no state tax states, there is at least a financial advantage. And due to the flexibility to found, get taxed and work in three different states, US companies can basically choose a state without additional income tax and operate from another one. Germany on the other side offers a tax-free level of up to 10,000 $ annual income (after having deducted the mentioned expenses above). Especially for founders with a small income this is very convenient and attractive.

Summing up, income tax from an entrepreneur's view do not really provide a major advantage in one country in the founding stage, because taxes have to be paid on both sides and are not the decisive element in the beginnings. The pro and cons for each country also balances each other. And since high personal earnings are usually coming much later after the founding in an entrepreneur's life, more income tax affects the founder much later. Therefore one can derive that an entrepreneur would not start a company and operate in a country or state, because they know that if they succeed they will pay lower income tax. It is very common among entrepreneurs to solve that issue with peers, tax consultants, investors or lawyers by moving that tax decision to a later point in time when success potentially comes closer, and since there is still enough time to think of a tax optimization and change of location. Before, however, there are many other important things to solve.

For personal investors and business angels that need to pay the highest tax rate, a lower income tax in the US between 3-10% than in Germany – depending on the state tax – could definitely help to build up higher fortunes and thus having more cash available for risk financing. However, as soon as investors and business angels run most of their money through corporations, the corporate and capital gains tax becomes the relevant tax.

5.9.2 Corporate Tax

Analog to the income tax, the corporate tax is the tax on the income of legal persons. Again, in the United States there is a federal, as well as, a state tax, while the basic *Körperschaftssteuer* ('entity tax') in Germany is only collected by the federal government. However, local governments in Germany collect the so-called *Gewerbesteuer* ('business operation tax'). This tax "basically is an additional corporate tax" (Zenthöfer & Leben 2008, p.133). Therefore, it can be considered as a local corporate tax and thus be compared to the different local and state taxes in the US.

5.9.2.1 Federal Concept and Comparison

The basic concept is very similar to the individual income tax. Every company is fully taxable as long as the business or the managing board is in Germany/ the US (Department of the Treasury 2006, p.2; Haas 2002, p.4) – this is another case, which this thesis does not consider. And of course the basic corporate income will be reduced by several exemptions, before it is subject to corporate tax rate (Department of the Treasury 2006, pp.9–17; Haas 2002, p.14).

Corporate Tax rate in Germany

For all generated profits the following federal taxes apply in Germany making the federal corporate tax very easy (Bundesministerium für Finanzen 2012):

- Corporate Tax: 15%
- Solidarity surcharge: 5,5 % of the tax amount
- Effectively: 15.825 %

Corporate Tax rates in the US:

US corporate tax is more complicated and much higher than in Germany. The lowest value is quasi the highest value in Germany. In addition, the rate increases above 35% twice. Companies between \$100,000-\$335,000 are subject to a 39% tax rate, where as companies between \$15,000,000-\$18,333,333 are subject to a 38% tax rate and is part of a more complex contemplation of receiving a higher effective tax rate (Keightley & Sherlock 2014, p.2).

$-Range	%-Range	Ascent linear...
0-50,000	15	Proportional
50,001-75,000	25	Proportional
75,001-100,000	34	Proportional
100,001-335,000	39	Proportional
335,001-10,000,000	34	Proportional

10,000,001-15,000,000	35	Proportional
15,000,001-18,333,333	38	Proportional
18,333,334- ∞	35	Proportional

Table 28: Corporate Tax Rates in the US (Tax Foundation 2011a) (Department of the Treasury 2006, p.17)

5.9.2.2 State Concept and Comparison

While Germany has a strict tax rate in the federal system, it offers local communities to determine their rate independently. It is not the state, but a municipality that defines a rate, thus even in a state there can be some significant differences. All companies, even small partnerships, have to pay this *Gewerbesteuer* as long as they work in commercial operations, which is basically every company interested in financial success. The rough calculation of the final tax of the *Gewerbesteuer* can be seen below, although the actual complete calculation is more complicated in reality (Zenthöfer & Leben 2008, p.136):

Basic income reduced by several deductions incl. 24,500 € tax free sum

= Subject to tax

multiplied by basic federal rate (3.5 %)

= Measuring subject

multiplied by local rate of assessment

= Taxes to be paid

Selected rates in Germany

Local tax	Local	Effective rate
City of Karlsruhe	410%	14.35%
State of Baden Württemberg ø	358%	12.53%
State of Bavaria ø	368%	12.88%
City/ State of Berlin	410%	14.35%
State of Mecklenburg-Hither	345%	12.08%
City/ State of Hamburg	470%	16.45%
State of Thuringia ø	349%	12.22%
Germany ø	390%	13.65%

Table 29: Selected Rates of Assessment in Germany (Author, (Destatis 2005))

In the United States corporate taxation works exactly the same way as shown by the individual income taxation. On the state level, there are also states that have not corporate tax like Nevada, South Dakota, Texas and Wyoming (Tax Foundation 2011b). In regard to the income tax, California (8.84%), Delaware (8.7%) and Massachusetts (8.8%) are above the average.

The differences in federal corporate taxes are quite easily shown. The Federal German tax system is much less complex and has lower tax rates. However, this advantage can be turned around when it comes to the local taxation system. Besides the very complex calculation and the high rates in Germany, the group of taxed companies is also extended with the *Gewerbesteuer* to small firms and thus affecting severely the startup group. The disadvantage in Germany in further increased, due to the danger of double-taxation, which will be explained in one of the following sections. For companies with less than 50,000 USD profits, which is rather typical for a startup or small company, the USA offers a lower total corporate tax rate. However, for profits over 50,000 USD the German total corporate tax rate would be lower at an average of approximately 28% (federal tax plus *Gerwerbesteuer*), if company in the US picks a state with a state tax of more than 3% e.g. California with 8.84%. But since many startups do not make any profits at all in the first years, the corporate tax rate should not have an effect on entrepreneurship from the founder's side. Nevertheless, the higher tax rate in the US for higher profits could have an effect on rich individuals and corporations, if they had less money to spend on startups. However, in each country there are complex forms of deductions on investments into startups and other companies, often connected with the capital gains tax.

5.9.3   Capital Gains Tax

A third type of taxes is called capital gains tax, which is collected in both countries only from the federal government. In general, capital gains tax has to be paid by a private person, who has gained income from its capital, such as real estate, and equity shares, such as stock sales and partly dividends. That is why this tax has a direct influence on investors, because it concerns also their revenues from investments. Since the systems and especially the tax rates in Germany and the United States are constantly changing, it is only possible to give a snapshot of the capital gains tax in the following section.

5.9.3.1   German Capital Gains Tax

The German *Abgeltungssteuer* ('flat rate withholding tax') is to be paid on incomes from capital, dividends, disposition etc. (Haas 2008, p.87). In general, the tax rate is 25 percent of the profits. However, one can choose to pay the individual income tax rate if it is lower.

Furthermore, there is another possibility called the *Teileinkünfteverfahren* ('partial income treatment'). Thereby an individual could pay the personal high income tax rate for 60 percent of the income with no additional rate for the remaining 40%. Requirements for this tax treatment are the following:

For dividends, shareholders must have an equity share of at least 25 percent or they must have at least a one percent share and be additionally an employee of the company (Bundesministerium der Justiz und für Verbraucherschutz 2009, para 32d).

In case of a sale, the shareholder must at least sell a one percent share of a company (Bundesministerium der Justiz und für Verbraucherschutz 2009 § 17).

A specialty for investment cases is the free flow equity with less than 10% shares according to § 8b Sec. 2 and 3 of the German KStG (Bundesministerium für Justiz 2014). Investors in Germany are able to pay very low or no taxes if their gains were reinvested in a "rollover" to another small company. This is very interesting for business angels, who often hold less than 10% in a company and reinvest the money they made from selling equity from their prior investments. Most new investors and wealthy family offices that have a classic industry background are unfortunately not aware of such benefits, when investing in startups. This could be seen as part of the taxation complexity, which needs refined and in parts more simplicity. The total issue is, however, very complex and must to be scrutinized by tax experts in a separate research paper. In addition, the tax situation is subject to change, because the German government is constantly planning on changing this exception (BAND 2013).

5.9.3.2 United States Capital Gains Tax

The American Internal Revenue Service (IRS) distinguishes between two different capital gains, the short and the long-term capital gains. The long-term gains are the only ones eligible for a tax deduction, because the short-term gains will be taxed through the regular income tax. Short-term tax applies to the increase of the value asset hold less than one year; long-term for value asset hold longer than one year (Turbotax 2013).

Ordinary Income Tax Rate	Short-term Capital Gains Tax Rate	Long-term Capital Gains Tax Rate
10 %	10 %	0 %
15 %	15 %	0 %
25 %	25 %	15 %
28 %	28 %	15 %
33 %	33 %	15 %
35 %	35 %	15 %

Table 30: Capital Gains Tax in the US 2011 (Tax Foundation 2013)

Ordinary Income Tax Rate	Short-term Capital Gains Tax Rate	Long-term Capital Gains Tax Rate
15 %	15 %	10 %
28 %	28 %	20 %
31 %	31 %	20 %
36 %	36 %	20 %
39,3 %	39,3 %	20 %

Table 31: Capital Gains Tax in the US 2013 (IRS 2013)

5.9.3.3 Differences and Potential Entrepreneurial Impact

For short-term period gains, the German tax system is the more favorable than the US, for persons with personal incomes leading to a tax rate above 25%. This could actually apply to almost all investors, which are by definition wealthy persons that have either money left over to invest or that work for a fund that not only pays high management fees but also huge bonuses in case of success cases. It must be remembered that investing in startups means not trading shares on a daily or a short-term base like on the stock market, thus the taxes on long-term capital gains is more interesting for business angels, as well as fund managers and investors. And for the long-term gains, investors are better off in the US, especially before the year 2013, when investing in shares and living from long-term capital gains increases had been very lucrative for rich individuals with a tax rate of just 15%. This led to a famous statement by Warren Buffet, who said that he is having a lower tax rate than his secretary, because most of his earning certainly came from capital gains and not a huge salary. Even with a planned change of the long-term capital gains tax in 2013 to 20%, it is still a very low tax rate in comparison to the income tax and lower than the 25% in Germany. In general, America has always been an investor friendly nation, and due to the lower benefit standards and payments of the social system, people had to secure their own pension well. That is why many Americans, even not so rich, are used to invest in the stock market and companies that they believe will guarantee growth and dividends in the future. All this is an ideal ground for venture funding, because people are much more open for investing in companies as part of securing their future income and pension. These considerations definitely increase the density of investors, which is often named as one important fact of the success of the Silicon Valley (Maier 2011a, p.9).

A positive regulation, however, is the rollover of free flow equity in Germany. As long as it exists, it is a tax advantage for investors in Germany, because such a regulation has not been found by the author for US free flow equity. However, the US has over proportionally more business angels than Germany, indicating that this effect did not really make up for the other differences (compare Table 7). Nevertheless, the US tax system offers, despite its complexity, attractive models of tax benefits in certain cases. It would require an expert team to analyze and compare the regulations with the German system.

An example for an additional tax deduction in this very complex tax situation is the possibility for companies to write off the already paid income taxes of individuals from their corporate tax. This is of interest for entrepreneurs, who own a major share of their company and can use tax loops to shift the taxes, thus reducing the total amount of taxes paid – privately and from the corporation. One of the most recent examples, is Facebook founder Mark Zuckerberg, who is paying high personal income taxes and at the same time reducing the tax payments of Facebook nearly to zero for many years (Kocieniewski 2012). This example of a very complex double-taxation case, shows the benefit Zuckerberg had. It must be noted that often double-taxation leads to problems and inequalities.

5.9.4   Double-Taxation Problem

Double-taxation is a problem occurring quite commonly in Germany. When a corporation pays part of its profits as a dividend to its shareholders, the company needs to pay corporate tax on the profit and the shareholder recipients of the dividends must pay income tax. Many countries try to avoid this double taxation by allowing the deduction of parts or the whole corporate tax from the income tax or simple by having no double taxation. In the United States, to an extent, it is possible to write off the corporate tax from the private income tax – thus possibly bypass some double-taxation.

The higher tax rates in Germany cause clearly a disadvantage for investors – from a financial point of view with this taxation issue even worst.

5.9.5   Summary on Taxation

The entrepreneurial (dis-)advantages are summarized in the table below. A "+" indicates a more investor or entrepreneurial friendly tax situation, where as a "-" indicates a more negative situation for one or both parties:

Germany	Issue	United States
➕ ➖	Complexity of taxation system	➕ ➖
➖	Danger of 'double taxation'	➕
➕	Corporate tax	➖
➖	Capital gains tax	➕
➖	Total taxation of wealth	➕

Table 32: Entrepreneurial (Dis-)Advantages of Taxation Systems

To sum up, the US offers a more investment friendly environment from a tax point of view, leaving more money for the investors, business angels, entrepreneurs and funds to invest, in order to avoid a double-taxation. On the other hand, the structure of the capital gains tax encourages people to invest their money – whether it is stocks from established companies or startups. Ultimately, there are more business ventures, due to more funding – thanks to in United States tax benefits.

Nevertheless, taxes are not the only reason, why America has more startups, more risk capital and more success cases. They definitely fill up the "tanks of money" for investments, but even more important than that, in the US there are many role model entrepreneurs and super successful founders like Bill Gates, Larry Ellison, Mark Zuckerberg, Larry Page or Herbert Kelleher, who have been successful, regardless of lower taxes, because of their drive, creativity and determination. Their biographies describe the passion for their concept that made them start and proceed in their living area with the best available resources. Especially the high taxes in California show that the tax aspect is not holding new high-tech and IT founders back. Unfortunately, international tax loopholes and maneuvers give successful corporations and their founders the possibility to avoid many corporate taxes and allow them to make huge profits outside their country of origin. The role model companies like Google, Amazon and Apple are also front-runners in this area (Reuters 2013a) a problem of international scale, which is of course beyond this thesis' scope. However, it supports the statement that tax aspects are not a necessarily hindering argument for entrepreneurs to start or develop their company.

5.10 Bankruptcy Laws

While taxes accompany the upswing of a company, one also needs to consider the influence and factors of the downswing side resulting in failure and bankruptcy, with the latter one being the ultimate from of failure at least from an entrepreneur's point of view. And a comparison between the legal handling of bankruptcy in the US and Germany indeed reveals major differences. The US bankruptcy laws are, according to the Federal Constitution (Art. I, Sec 8), to be regulated only by the Federal Government. The laws are not only to apply and to give protection in all states equally, but also give indirectly natural and legal persons a second chance. Thus, it became possible for entrepreneurs to fail and to be protected more than once in business; thereby opening the door for future success. This is coherent with the stereotype of US entrepreneurs failing several times till they finally become successful. When filing for chapter 7 under the US Bankruptcy Law, the insolvent entrepreneur as an individual may be discharged of all liabilities – in most cases within a few months – and thus has a chance for a 'fresh start' (US Congress n.d., p.Chapter 7; Gov UK 2012; Mathur 2011, p.9).

The discharge under Chapter 7 can be denied, if a person has filed for example for Chapter 7 within 8 years before. Chapter 7 applies for individuals and corporations; the discharge however is for individuals only. However, entrepreneurs, who have failed with their company, can file for bankruptcy for the company and for themselves and then free themselves from the liabilities tied to the company. Certainly, all real and personal property needs to be listed and will be sold, if the company's liabilities exceed the property assets. But since, most entrepreneurs in their first ventures have not build up personal wealth, they can more easily "strip down" and then start again with a new venture or in an employee position. In Germany, the regular period for a discharge takes 6 years and a prior discharge in some cases is possible after 5 years (Insolvenzrecht.info 2012). There is a major difference of bankrupt perception between the US and Germany, which underlines the fear and stigma of German entrepreneurs. After failing with a business in Germany it is very hard to recover, and the personal damage being insolvent and controlled by an insolvency trustee is devastating (Europolitan 2006). The insolvency trustee will use almost all income from the entrepreneur in that 6-year period to pay back the liabilities, and most personal property, which is not essential for living, will also be taken away. The details and legal interpretation in each country, however, require legal experts, in order to further compare both bankruptcy regulations.

5.11 Gender Aspect

A frequently discussed issue in entrepreneurship support is the gender aspect, because starting a business traditionally was, and still is, a male domain. In the US, only 1 out of 3 entrepreneurs is a women (Kauffman Foundation 2011, pp.6–8). The numbers in Germany are quite similar according to the Global Entrepreneurship Monitor (Global Entrepreneurship Monitor 2012, p.12). However, other sources like the German Startup Monitor report of only 13% female founders, when it comes to compare founding a startup vs. to being self-employed, without having any employees (Deutscher Startup Monitor 2013). 13% or 325 out of 2,500 is the number of females having received the federal German scholarship EXIST *Gründerstipendium* since 2007, which suppose to support innovative startups with high growth potential (Zeit-Online 2013). Most sources agree that the success rate and quality of a founder are not depending on gender, although some difference in performance and motivation occur leading to different outputs (Kepler, & Shane 2007, p.1). The behavior and attitude may play an important role, when it comes to business interaction or risk financing (also compare competition and gender on p.77). Men are said to be more number driven, risk taking by nature, self confident of their skills and talking more often about the potential of their own startup. Women on the other hand tend to talk about their motivation starting their own company, are usually not so number driven, less risk taking and talk less self confident about themselves (Caprino 2013). Nevertheless, the gender aspect does not seem to explain the reasons, why US startups perform better and are more successful than other international teams.

5.12 Top 100 Established National Companies

When comparing the US and Germany it also helpful to observe and compare the best performing and largest companies in each country, in order to see, how far the entrepreneurship advantages in the US might have influenced the complete industry chain and economy.

At first sight, Germany and the US have a similar structure when it comes to their Top 100 companies (DAX, MDAX, SDAX and TecDax in Germany and Fortune 100 in the US). Both countries are still dominated by their Blue Chip Corporations, which have been in each country's index for decades. This is especially the case for the DAX and the Top 30 US companies. These Blue Chips come from classic industries like automotive, engineering, retail or banking, What Germany lacks are hardware computer companies like HP or IBM. Neither country index has a newcomer in its top ranks. In Germany the youngest company in the DAX is SAP, which is also nearly 40 years old (boerse.de 2014). However, the last two decades have brought forth an industry, which caused the most dramatic change in business and personal life – the Internet. All the startup companies in this field have been influenced by this new technology, and they brought forth innovation, as well as created many new jobs. Nonetheless, they still have not reached the top players in terms of revenues. The only startups that grew, within less than 15 years, into a prime Top 100 position are Google and Amazon (Fortune 2014). Aside from the US *heavy weights*, Google and Amazon, the established companies like Apple, Microsoft or Cisco are playing a much more important role in this new digital and Internet world. They are offering services in these fields, interacting more with the startups, influencing trends on their own and adapting to them quicker.

One level below the top 100 companies interesting difference between Germany and the US became visible. The SMEs in Germany are still following the path of the classic industries (Compamedia 2013), linking themselves strongly to the German Blue Chips as their suppliers and thereby become globally successful. Germans are proud of their hidden champions – midsize companies that are often world market leaders – without having a high general visibility. But when it comes to these "new industries", there is a lack of global impact – especially in high performing IT, software and Internet companies. Lists of the most influencing online companies simply do not include German companies (Anderson 2013; Anon n.d.; Pozin 2012). The same can be seen for biotechnology and medical companies, which are in Germany either highly dominated by large corporations especially pharmaceutical ones or not covered at all (Marshall 2008; Ranade 2008).

Aside from the energy sector, Germany does not play a major role in these important new industries. In both countries more than 50% of their VC money is invested in Internet, Software, IT or mobile startups. But since the general VC volume in Germany is so much lower, Germany does not receive a critical mass of investment in these areas, in order to play a significant role. Due to this lack, Germany does not produce global market leaders in these business segments (PWC & NVA 2013; Majunke Consulting 2013; Centre for Strategy & Evaluation Service 2012). However, not only because of the low investment size, but also because the Internet – including the mobile Internet – is influencing business and personal life in every facet, its importance and effects are more visible and immanent in the startup and innovation world than biotech, cleantech or other industries, thus leaving Germany behind in the limelight of the startup world. These differences and focus on old industries in Germany while only following the US players in new technology areas like IT can best be described by the phenomenon of US *copycats* in Germany.

5.13 Internet Success Cases & Copycats

When analyzing the Internet and IT scene in Germany, there are few "lighthouse startups" and success cases that have been innovated by Germans and that have influenced the world market. Besides companies like the gaming producers BigPoint and Gameforge most other known startup cases are so called *copycats* of US startups.

The following table displays some of the most prominent cases with the US originals all being billion dollar companies in evaluation:

US startup (launched in)	Evaluation/ Market Cap. 2013/2014	Profit 2012/ 2013	Online concept User or sales	German copycat example
Ebay (1995)	$67bn (07.01.2014)[7]	Yes	Auction place for End-consumer	- Ricardo (merged to QXL Ricardo in 2000 and finally sold to a Naspers for $1.8bn in 2008) - Alando (sold to ebay for $54m in1999)

[7] (Yahoo Finance 2014), Market capitalization: Ebay, Amazon, LinkedIn, Facebook, Yelp, Groupon

Amazon (1995)	$180bn (07.01.2014)[7]	Yes	Marketplace initially for books	- Buch.de (Market Cap. $140m on 07.01.2014)
Zappos (1999)	$1.2bn (sold to Amazon in 2009)	n.a.	Marketplace for shoes	- Zalando (Eva. 2013: $4.6bn, not profitable) - Mirapodo (part of Otto group)
LinkedIn (2003)	$24,4bn (07.01.2014)[7]	Yes	Social networking platform for professionals	- Xing (before OpenBC) ($595m on 07.01.2014)
Airbnb (2008)	$2.5bn (Eva. 2013)	Yes	Platform for private offering of homes and	- 9flats - Wimdu
Facebook (2004)	$140bn (07.01.2014)[7]	Yes	Social networking platform	- StudiVZ (sold to Holtzbrink for $110m in 2007)
Youtube (2005)	$1.6bn (sold to Google in 2006)	n.a.	Video sharing platform	- MyVideo (sold to ProSieben in 2006 & 2007 for app. $50-70m)
Yelp (2004)	$4,7bn (07.01.2014)[7]	No	Service for comments & evaluation on places/ stores etc.	- Qype (sold to Yelp for $50m in 2012)
Etsy (2008)	$0,6-1bn (Eva. 2013)	Yes	Marketplace for handmade items	- Dawanda (Eva. 2012 ~$25m)
Groupon (2008)	$7,9bn (07.01.2014)[7]	No	Platform for local special deals	- City Deal (sold to Groupon for $126m in 2010) - Daily Deal (sold to Google for $118m in 2011, rebought by founders in 2013)

Table 33: German Copycats and their US Origin

Supporting Sources: (Yahoo Finance 2014; Slavet 2013; Forbes 2013a; Deutsche-Startups.de 2013b; Techcrunch 2009; Telegraph 2011; Techcrunch 2011; Techcrunch 2012; Strauss 2012; Boldt 2011; Cloer 2012; Hoffmann 2013; Kaczmarek 2011)

All copycats have not been more successful than their US origin, and rather tried to be bought by them or a strategic competitor, rather than overtake and lead the world market itself. The valuation of the US companies in these cases ranges between 25 to more than 1000 times higher than of its German counterparts. The only exception is Zalando, which ranges above the original Zappos evaluation, when it was purchased by Amazon in 2009. However, insiders say that the original intention of getting bought by

Amazon much earlier did not work out, thus the founders and investors of Zalando needed to continue to grow the business and invested more than 250m € in venture money (Internetworld 2013). Zalando shifted from being a selling point for shoes to become a much wider marketplace for other products. It has become the number three web-shops in Germany, after Amazon and Otto (Handelsblatt 2014a). Zalando is professionally financed and advised by the founders of Alando, the Samwer brothers. They have professionalized the copying of US startups and are currently Germany's top Internet entrepreneurs and the only one with a real global perspective. Aside from Yelp and Groupon, all US companies, mentioned in the display above, are profitable, still growing and heavily investing. Even the acquired and integrated companies like Youtube and Zappos are said to be profitable for their acquirer according to unofficial sources.

US companies not only inspired German entrepreneurs to copy them but also those of other nations. This underlines the influence these startups can have, and it is a good indicator of the importance of a startup on a global scale. German startups yet lack such role models or world influence – not to mention leadership.

When considering further success cases of startups in the last 10-15 years, as well as companies being mentioned in the media for being the most influential or successful startups or companies, two aspects become very obvious (Deutsche-Startups.de 2013a; Poets & Quants 2013; RedHerring 2013; Pozin 2012):

- Most successful companies in that period are Internet and IT companies incl. mobile (>60%)
- The largest number and the most successful companies are coming from the US

This shows that Table 33 above is not only talking about ordinary cases but rather about the core of the current startup scene and the dilemma of Germany, because it is not playing a decisive role in the new industry segments, especially of Internet and IT, rather is still focuses too much on classic industries such as machinery, automotive or chemistry. With the new mobile industry on the rise outgrowing its parent industry – the Internet, it is obvious that Germany is not influencing this market as well. Besides the dominance of the smartphone companies like Samsung from South Korea, Motorola and Apple from the US and the new joint company of Microsoft and Nokia, it is also software, application and gaming firms that seem to have golden times ahead of them. Unfortunately for Germany, it has no global player in this hardware or software field – aside from BigPoint, which is, however, primarily owned by US private equity funds (Bloomberg 2011).

Entrepreneur and so called business guru Anderson was supporting theses facts, when he came up with a list of the 30 most important online companies. Not one came from

Germany and only one from Europe – Skype – which was mainly financed by Draper, a US venture capital company (Anderson 2013). The following list describes some of these important online companies that are dominating their business sector and had been mentioned by Anderson as well as by the author himself:

Internet search

- Google with its founders Larry Page and Sergei Brin originated in Stanford at a computer science course, where Page and Brin had been inspired and encouraged by their professors to continue the examination of the mathematical properties of the Internet and World Wide Web. They developed a special algorithm, which revolutionized search functionalities in the Internet and swept aside companies like Yahoo and AltaVista, which had been dominating the search market. Its influence and further development in various fields of IT and business cannot be overvalued.

Online retail

- Amazon started with the desire of its founder Jeff Bezos to participate in the vast projected growth of the Internet. Being a Wall Street trader in the computer science field, Bezos tackled the concept from an opportunity side, narrowing down the most interesting products to sell online. Between New York and Seattle it was said that he founded the company. Today, Amazon is the largest online retailer in the world.

Social network

- Facebook started at Harvard when Mark Zuckerberg wanted to create a network to connect people within his college and to rate girls on campus. Facebook grew to the largest social network in the world with more than one billion users engaging themselves in private and business affaires.

Online payment

- PayPal had been originally designed for online payment with Palm Pilots when its founders Peter Thiel, Luke Nosek and Max Levchin merged the company with X.com. opening it to for general e-commerce payment in many areas and on many devices. It is today the largest online payment method besides classical credit card and bank transactions.

Functionality applications & entertainment platform

- iTunes was created by Jeff Robbin, who created a device which was able to play music form an MP3 file. Apple bought the technology and renamed it iTunes and developed it to a powerful centerpiece of its business. It is used today to steer the end-user's complete Apple product portfolio (iPad, iPhone, iPod, MacBook), from online registration to providing the user with all necessary applications and content (music, video, gaming and other programs). iTunes, the pioneer on its field, inspired other platforms, such as Google playstore, to follow.

Cloud storage

- Dropbox was one of the first companies to provide free online storage using so-called cloud technology. Content can be shared simultaneously between users and is being saved on the hard-drive as well as in the cloud. Big players like Google or Microsoft have followed with their Google Drive or SkyDrive.

Online travel agency

- Expedia originated at Microsoft and had been developed by its founder Mark Schroeder to have a platform for many travel services. Today, it is one of the biggest travel-booking site on the Internet. It is said to be one of the first companies that has cleverly been using marketing and social media engagement to outperform its competitors in many countries.

Video platform

- YouTube had been created by Chad Hurley, Steve Chen, and Jawed Karim as a social place to share online videos and visual content. YouTube is owned since 2006 by Google and has the most visitors and content in the world for videos online.

The examples above are all fairly new companies, which demonstrate the movement "From startup to global player" category. Not one of them is older than 20 years, in light of the fact that they all have been created in the Internet age. Nevertheless, established companies like Apple, Microsoft and Cisco also have taken an important role in this Internet age and must be mentioned when talking about the most influential companies in the Internet.

As the Internet and the mobile industry have such a strong influence on other business segments and entire industries, it is a dangerous shortcoming that Germany has to make up for. Google is a classical example for this significant influence. Moving forward from an end-consumer oriented search engine to a business search engine in the early years. It has, since the establishment of its search algorithm, conquered many diverse fields. In the beginning, it extended its business to all Internet and computer relevant areas, such as sharing documents, email communication or map routing. Further expansions then included video, social and consumer platform. In acquiring Motorola it now seriously entered the hardware and mobile market. The *Googlecar* opened a new field towards the next generation of automotive. Its latest major acquisition of Nest on January 13th 2014, demonstrates the conquering of further relevant areas of consumer and business life, combining home, energy and Internet (Bloomberg 2014). According to the English Wikipedia site, Google has acquired more than 140 companies in the last 13 years (Wikipedia 2014c).

Even when comparing non Internet and IT startups that are top funded or well performing, the situation is almost the same: America is leading the way, leaving Europe and thus also Germany behind.

Of course the VC funding plays also a major role in these cases. What is crucial for most Internet and IT startups is even more essential for high-tech startups. The necessary investment sums for high-tech companies initially is much higher and revenues and profits usually come in much later than shown, for example, by the Internet cases. And due to the much higher VC volume in the US, the chances of getting funded in the US or by a US VC fund are far greater.

Since this chapter is comparing Germany and the US in terms of their startup bases and traits, one has to ask, why all these success cases are coming from the US and not from Germany, although Germany is a rich, well-educated high-tech nation; and the fairly young Internet, as well as IT industries do not depend on many domestic natural resources. The same is true for many high-tech cases, for which mainly topnotch universities, high fund volumes and research friendly laws and regulations are necessary. It could have been a great success story for Germany, to have conquered this industry and to have led from day one. Especially when realizing that key technologies like the World Wide Web or the mp3 player had originally been developed in Europe or Germany (Businessweek 2007).

5.14 Top 100 Global Startups

The media agency Red Herring, also known for its innovation magazine, publishes out a yearly list of the Global Top 100 startup companies. This list has become an international well-accepted source for promising new companies and entrepreneurs. The Red Herring agency was among the first to identify companies such as Google, Facebook or Skype (Paymetric 2013), but it does not focus on online and digital companies. When comparing the years 2011 to 2013, the US with a total of 117 nominees clearly outperforms Germany with only 8. This is another unfortunate indicator for Germany, showing that Germany is not on the forefront of current entrepreneurial endeavors. The complete list can be found in the **Appendix C.**

5.15 Summary of the US-German Market Comparison

The comparison between the US and Germany (to an extent also Europe) on their differences in the entrepreneurial setup and success, has covered many areas. The following list will summarize some of the most important areas and their characteristics (US vs. Germany):

1) Risk Capital: high volumes & many investors vs. small volumes & few investors
2) History: "Bold genes" & US (Anglo-Saxon) tradition vs. German continuous heterogeneity and theoretical tradition
3) Geography: Conquering new territory & staying isolated vs. continuous European mingle and wars
4) Culture & Society: American Dream vs. Germany envy
5) Market: Homogeneous huge market & one language vs. heterogeneous, fragmented markets & multiple languages
6) Government regulations: Fresh start & low taxes vs. years of insolvency & high taxes
7) Character Traits: Opportunity seeking vs. risk aversion
8) Top performers: IT & High-tech vs. classic industries & copycats

The comparison has shown many advantages and reasons, why the US venture- and entrepreneurship scene is far more advanced and successful than the German one. And even if Germany was part of a unified Europe, the advantages might still be on the American side for quite a while.

These gained insights and results will help to define a holistic Entrepreneurial Ecosystem in the following and are the base for improvement in these areas.

6 The Entrepreneurial Ecosystem

In the previous chapters, the framework of entrepreneurship and the success factors of entrepreneurs have been collected and analyzed comparing specifically the German and US markets. These factors either come out of or are influenced by the Entrepreneurial Ecosystem. With the idea of better defining and improving this ecosystem to foster entrepreneurship and innovation, one should especially concentrate on the factors that do have a strong influence on startups' success and bring them into the context of an Entrepreneurial Ecosystem.

6.1 The Basic Elements

What this thesis plans to do is to build the ecosystem directly around the entrepreneur, with different levels of proximity and importance for him or her. This approach will be not only inspired by Isenberg and his six domains (Markets, Policy, Finance, Culture, Supports, Human Capital) but also by Timmon's process of Entrepreneurship (compare 3.5 on Entrepreneurial Ecosystems).

6.2 The Ecosystem Circles

At the heart of the of the entrepreneurial process is the founder: The seeker, creator, and initiator behind the start-up (Timmons & Spinelli 2004, p.1).

This new Entrepreneurial Ecosystem model puts the entrepreneur in the center and develops the areas or domains, in form of circles that develop from the in- to the outside. The circles almost represent a chronological order, when the entrepreneur encounters the different elements – starting with his family, to education and up to the business world.

The areas are systems of their own and together form the Entrepreneurial Ecosystem, with its own rules and elements. As for the influence the domains have on the centered entrepreneur, one has to distinguish between indirect and direct effects.

The **indirect influences** are:

- Subconscious influences
- Frameworks like culture and language

The subconscious ones constantly shape the awareness, self-understanding and desire to become and not to become an entrepreneur. They later continue to inspire and play a role during decision making in many different situations. Frameworks like culture might have similar effects, however together with the language they can make it easier or more difficult to develop and communicate ideas. The Greek language for example was perfect for philosophical thinking, because of its clear and sophisticated structure and abstract forms like the "it" (Babinotis n.d.).The English language is spoken around the world and helps to break barriers with its informal "you". And especially the American culture encourages an easy get-together, which is of great advantage for entrepreneurs to make contacts and get things started.

The **direct influences** are a set of features, of which the most important ones are

- Advice
- Financial support
- Contacts
- Guidelines & templates
- Market access
- Legal & tax issues

The direct factors have not only an impact on the entrepreneur, but also on the other elements in the Entrepreneurial Ecosystem. That is why many elements within the Entrepreneurial Ecosystem are not independent from each other. Through certain tax reform for example investors might be able to collect more risk capital or open new funds, from which the entrepreneurs will profit. A positive entrepreneurial culture is another example, which will engage established companies or customers in general to accept and work also with startups or buy their products.

Social online networks are a special quite new form and influence in the entrepreneurial world. They also have indirect and direct impacts. On the one hand, they are a part of life for many especially young people and entrepreneurs – thus of their inner circles (see next sections). On the other hand, they are used for business purposes and contacts, as well as to build an entrepreneurial system of their own – from new venture potential to customer acquisition. Thus, they could be considered to be an element, as well as a set of tools.

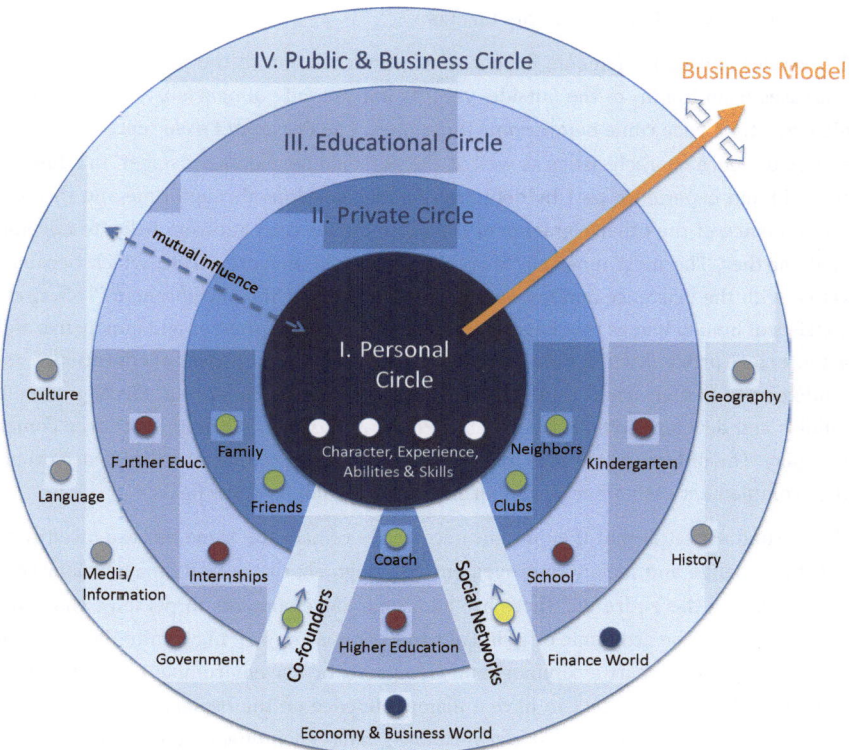

Figure 24: Entrepreneurial Ecosystem

- The circles represent areas of influence on the entrepreneur that are arranged on the one hand in a chronological order, in which they get in contact or are part of the entrepreneur. On the other hand the circles also represent the proximity to entrepreneur as a person, moving from close and personal to the outside of Circle III and IV
- The dots stand for the influencing elements, which have their origin in the according circle
- All circles and elements are connected to each other and can influence each other
- The white dots are about characteristics of the entrepreneur
- The green dots are about persons
- The red dots are about institutions and companies that offer the services and set up rules
- The grey dots stands for framework elements
- The blue dots are elements as well as a new ecosystem, which have great influence on the entrepreneur
- The yellow dot represents the social network element, which is a mixture of several areas like people, framework or ecosystem itself
- The business model bar reaches over all circles indicating the influence by all circles and elements or the model. The arrows at the bar underline that constant movement and development of the business model and thus the different touch points with the elements at certain times.
- The two corridors left and right from the Economy & Business World stand for the connection of all circles and their elements. The co-founder and social network element can also move between the II, III and IV Circle indicated by the arrows

6.3   The Laws and Linkage of the Circles

The circles and their elements have a strong time relation to the founder's life. Time increases from the in- to the outside, yet it is not the only dimension, and some outer elements time wise come before some inner ones. The *Personal Circle* represents more or less the born characteristics as well as the personal development stages. In addition, the skill and experience part by definition further develops through time and through the influence of the other elements of the other circles e.g. education helps to develop skills further. The most inner circle is followed by the *Private Circle*, which becomes active with the existence and increasing consciousness of the entrepreneur. This circle consists of human beings that have an influence on the entrepreneur and mostly live and act in a close proximity to the entrepreneur. With the different phases of education, from kindergarten to high school and potentially university, the *Educational Circle* gradually follows and develops. When touching the *Public & Business Circle* the last area comes into play. The difference of each circle indicated, that they cannot substitute each other; they complement each other by adding and opening complete new fields.

The second aspect is that they do not necessarily remain the same but are faced with constant change and ideally improvement. Although character traits may remain very similar during the entire life, there are additional developments in person's character over the years. The economic or political situation in a country may ignite the desire in an entrepreneur to start a company and change his or her environment and life. Thus, the *Public & Business Circle* can have a major influence on the *Personal Circle*. It is also likely that these environmental circumstances have an influence on the *Private and Educational Circle*, thus also influencing the founder from these two levels. For the circles two, three and four however, it is more obvious to understand these constant changes. Entrepreneurs will make new friends, have several coaches, co-founders may reach their Personal Circle as well as a constant flow of social network contacts are influencing this circle.

The *Educational Circle* itself is spread over a long period of time anyway, and may continue in form of a life long learning process, as well as in form of a coach or mentor, who from out of the private circle can influence and teach the founder. The *Public & Business Circle* certainly changes through out the entrepreneur's and the company's time with each new customer, employee or economic and political development.

Another perspective or dimension for the setup of the circles is personal closeness of the areas and their elements to the entrepreneur. The *Personal Circle* is the closest, because its elements are part of the entrepreneur. The *Personal Circle*, as the name reveals, stands very close to the entrepreneur. The *Educational* and the *Public & Business Circle* of course stand in a greater personal distance to the entrepreneur.

The basic law of the ecosystem is that elements of each circle remain in their circle, but can influence and possible change each other through various ways. The three other circles and their elements all have a direct or indirect influence on Circle I, with the character traits of the entrepreneur. It is possible to add and remove elements from the circles and give them an influencing, impacting factor, as well as new personal or other aspects and relations, which goes however beyond the scope of this thesis. When applying this ecosystem to other countries, some changes might be necessary to adapt it to the relevant aspects of time and place. However, this Entrepreneurial Ecosystem is designed in such way that it suppose to allow a comparison between various countries without the necessity for change: This is due to the open or flexible design of the circles and elements offering enough possibilities to allocate new aspects to the existing elements. However, it does not claim to be a complete model and applicable for all types of entrepreneurial questions and situations. It shall rather be seen as a concept and template to be adapted over time.

The next chapter will discuss these circles and elements in detail and also take a closer look on their interaction.

6.4   The Personal Circle - Abilities, Skills, Experience and Character

The first area and most inner circle is the Personal Circle or Entrepreneur Circle. Like in any centered system, the main player and orientation point has to be understood, in order be able to interpret the relation and interaction between the other elements. Since an entrepreneur is a person, the understanding process needs to cover *abilities, skills, forms of character and experience*, which best describe a person. Abilities are pre-given competencies and, more or less, non-trainable and thus predisposing a person to be good at something e.g. creativity, learning aptitude or simply the ability to see (TheFreeDictionary 2012). In contrasts, skills are competencies to do something that has to be trained and can be improved (DAA-Stiftung n.d.), e.g. driving a car or creating a business case. Skills can be used to improve the pre-given competencies (abilities) for example being able to drive or being able to learn a language. However, a person that became blind up to today, is not capable of driving a car, no matter how good the driving skills had been before. All human beings have the ability to learn a language, but for some it is much easier than for others, which is then usually considered to be a talent.

The various individual types of character make up a complex physiological field. Whereas, the layman would think right away of character traits such as endurance, tolerance or humor, the expert will start much more general with five categories also known as "the big five" or "OCEAN", which have been introduced by Paul Costa and Robert McCrae (De Raad 1998). These stand for the prime factors of a person's character: openness, conscientiousness, extraversion, agreeableness and neuroticism (further details on *OCEAN* theory will be explained below at p. 157). Another very important and often named aspect about a person and their success is experience. It can on the one hand improve the skills and add knowledge, while on the other hand experience will also sharpen the senses and intuition (Oxford Dictionaries 2014).

Many papers have described and analyzed the characteristics of abilities, skill, character and experience, but have not connected them as being part of an ecosystem. Yet this is crucial, because if one adapts the Entrepreneurial Ecosystem to improve it, it has to rather fit the entrepreneur's conditions and not the ones of those, who do not participate or are not the key player. That does not mean, that society cannot place rules and set ground for moral and ethical beliefs. Quite the contrary, it actually has to do it, but not rules that inhibit economical progress and success. One has to keep in mind the perspective of the entrepreneur, in order to understand, what drives and motivates him or her. Entrepreneurs do things differently and want to set up their own rules. They have a strong desire for independence, which can lead to friction with society and governments, who need to set the rules for the social welfare and peace. The rising number and awareness of social entrepreneurship shows the openness and interest of entrepreneurs to do something good for society and not only work for their own benefit. Spinelli and Adams (2012, pp.223–225) also explain that social entrepreneurship does not exclude per se making profits while acting well for society. Therefore, it should be possible to set the ground for a better and more ethical business culture, in which society in total profits from entrepreneurship, while still leaving the entrepreneur freedom to prosper under his or her own terms.

6.4.1   Ability & Skills

Through the example of learning a language, abilities and skills can be very well explained. It is well known that the ability of learning a language varies from person to person – some humans learn a language much easier than others. One speaks of language skill, when a person learns a language and its application better through the use of a certain system. The application of the skill increases the ability to learn a language. Applying this idea of learning a language to entrepreneurship, one could say that at least every human has the ability to found a company. However, nearly establishing a company does not make someone an entrepreneur and does not make the founding a success. Mastering this newly created company requires additional abilities and skills, such as learning a language. Of course there needs also to be an interaction within the ecosystem for becoming a successful entrepreneur, which will be explained in the following chapters. Since the basic ability for learning a language or founding a company is generally rather low, the person is bound to have greater problems further down the road with many more aspects and difficulties lying ahead.

This can also be observed in music or sports. Everyone remembers having had classmates that had a great ability, often referred to as a talent, for a particular sport. While the less talented were still trying to master the basics, the talented persons already moved on to try out new things, to improve their game, to repeat certain movements and tactical elements. This fact continues when observing semi-professional sports, where a talented tennis player for example will play in a much higher league with the same amount of training. The higher the professional level the more hard work needs to be added for a successful game. However, the question remains, whether a non-talented player can ever compensate his or her lack of talent by working harder than the opponents (Hambrick & Meinz 2011; Anon 2014). Since the most talented persons can also through hard work excel, the less talented may only pass the least talented players. Transferring this fact to entrepreneurship, it may well be very difficult for a rather non-talented entrepreneur to start something new, compensating many of the issues without having the proper traits and strengths. Spending the time and hard work in an area, in which the person has its strengths, could be more effective and successful.

6.4.2 Character Traits

The surveys of many researchers have indicated that entrepreneurs differ in certain categories from non-entrepreneurs, and that certain factors might have a greater influence on success than others (compare p.14; p.80). Results from a recent study, which was published in the Administrative Science Quarterly in June 2013, show that CEOs and leaders with a strong tendency to narcissism lead their companies to more innovation (DIE ZEIT 2013). This underlines the hypothesis that character traits definitely have an influence on success and innovation. And therefore, the results also speak for the entrepreneurs, who are leaders of their companies. But again it is Prof. Faitin (Faitin 2008) in his book, "Kopf schlägt Kapital", contradicts this by mentioning studies in the US, which said it was not possible to correlate success of an entrepreneur to certain character traits. Maybe it is the approach of scientific thinking in the search for the one specific trait or the exact trait combination or the right gene, which could scientifically define success that explains this contradiction. It is not in the nature of science to accept that a general combination of character traits and other influences leading to success, cannot be defined to a full extend, like it is the case with the Heisenberg uncertainty principle of the position and movement of electrons, which keep on moving and never have the same position (Pirner 2007, p.4; Heisenberg 1927, p.176ff). Thus, the question arises, can an entrepreneur succeed without having certain character traits, which have been scientifically defined?

But it is the author's opinion – in contrast to others like Prof. Faitin (Faitin 2008, p.4,16) - that not everyone could and should become an entrepreneur. Faitin reasoning that a successful entrepreneur just needs to think his idea over and over again, in order to find and shape the right business concept is incorrect. Faitin believes that with the outsourcing of almost every field of a company, the entrepreneur needs to have fewer skills and abilities than expected, and thus many more persons could become entrepreneurs. However, thinking an idea over and over again, requires a certain mindset, intelligence and endurance, which are part of the character traits of an entrepreneur. During that process, there will be also a lot of doubts and problems, which could become very demotivating, thus more than ever certain type of character traits like confidence or self-motivation are needed. And a successful implementation always requires time, which leads to a gap of input and output. Many persons are afraid that the input – whether it is time or money – might never pay off. For the author, the group of potential entrepreneurs is, therefore, quite small, because on the one hand the risk propensity for an uncertain input-output relation applies only to a small group, and on the other hand the characteristics of self-motivation and endurance are something non trainable and occur only among a small group of persons. Hence, teaching entrepreneurship, and improving or training skills cannot generally expand the potential group of entrepreneurs. The definition of Stevenson (p. 20 oben), according to the author of this thesis, can and should also be used for defining the preconditions or characteristics for an entrepreneur:

- Self motivation & endurance
  (→ *the pursuit of opportunity beyond the resources that you currently control*)
- Dealing with a high personal risk propensity
  (→ *the pursuit of opportunity beyond the resources that you currently control*)

Nevertheless, among the group of potential entrepreneurs that match these preconditions, there might be a large number that never come forward, establish a company or become successful, because they lack the necessary skill set and support in their ecosystem. Still this group regardless how large and can be influenced to become an entrepreneur. A question though remains, whether the self-motivating characteristic by itself also leads to "taking action" and starting the venture or whether this requires additional or different characteristic. Taking action is certainly also part of a decision making process, which analytically can be trained and thus became a part of the skill requirement; however, the actual making of the decision – in this case the action to start a business – cannot be trained.

Zhao & Seibert (2006) were looking into that from a different angle, by comparing general characters traits among certain groups. They discovered by comparing 23 studies that entrepreneurs actually differ from for example managers in the *OCEAN* categorization of a character. They score higher in the field of conscientiousness, as well as openness and lower on neuroticism and agreeableness. The following listing will describe the *OCEAN* elements in the results of Zhao & Seibert and explain the relation to entrepreneurship. When *OCEAN* is used to describe a person, one reckons how strong a person scores in each category (Simmons 2009):

- **Openness**: Someone who is open, intellectually curious, creative, alert, imaginative, reflective, untraditional and can empathies with others. Entrepreneurs are **strong** on this trait.
- **Conscientiousness**: A reliable, efficient, decisive and pragmatic person, who is well organized and goal oriented. Entrepreneurs are also **strong** on this trait.
- **Extraversion**: An optimistic person with courage, who is persuasive and likes to mingle with people. Entrepreneurs vary in this discipline.
- **Agreeableness**: It reflects the interaction among humans. Scoring high means being trustful, forgiving, caring, and cooperative, while a low score stands for a manipulative, self-centered, suspicious and ruthless person. As entrepreneurs have to fight among their way to success, they also can **score weaker** on this trait according to Zhao & Seibert.
- **Neuroticism**: This trait describes how individuals experience negative emotions like anxiety, hostility, depression, impulsiveness and vulnerability. Scoring low on this trait stands for an emotional stable, self-confident, calm, even tempered, and relaxed person. Entrepreneurs need to **be strong** in this field and thus score low, in order to keep on moving also in bad times and not to lose faith.

6.4.3 Experience

Experience is the element that uses knowledge as well as intuition and grows with time (Timmons & Spinelli 2004, p.134). And with this growing experience, entrepreneurs will be able effectively apply their abilities, increase their skills and handle better many other elements of the circles of the Entrepreneurial Ecosystem. Even creativity and boldness, which sometimes might be blocked or slowed down by experience, can be nurtured, if the experience leads to more calm and patience, rather than fear and impatience.

6.4.4 Comparison of the Personal Key Elements

The following table will try to compare these characteristics with the success factors for entrepreneurs that have been derived from the surveys as well as from literature research.

Type	Potential Influence
Abilities - Base intelligence - Learning abilities - Creativity - Self-motivation - Quick learning - High energy level - "People mover"	Pre-given and further nurtured characteristics: - Base for any job - Base for learning anything - Making things different and potentially better - Keeping on pushing and continue in hard times - Increasing chances to improve in a short time period - Base for endurance and strength, influencing staff - Getting the optimum out of the personnel
Skills - Business skills (i.e. Accounting, Analyses, Business Development, Marketing, Project Management - Engineering, IT skills (i.e. Production, Programming, Research & Development, - Applying knowledge	Acquired and trained: - Doing the basic job of a businessperson and entrepreneur - Interpreting, planning and acting - Leading in a certain technology field - Being able to hire and to lead the right people in that particular field - Applying certain tasks on your own - Transforming theory into practice
Character - Open minded - "Doer" - Hard working - Endurance - Need for achievement - Integrity - Honesty	Pre-given characteristics how a person is: - Seeing & listening to opportunities & new ways - Starting something, getting things rolling - Getting things done, handling workloads - Finish something - Continuing effort - Building up and sustaining relationships - Credibility of employees, partners and clients
Experience - Business experience - Entrepreneurial experience - People experience	Increasing knowledge and intuition: - Deal with clients, understand market developments, manage and evaluate businesses - Start and run a business, overcome hurdles, evaluate situations, stay calm and focussed - Motivate people, balance a team and pick the right people

Table 34: Personal Traits in Circle I

The Personal Circle with its natural abilities and the trained skills also sets ground for the interaction with all the other circles and elements of the ecosystem. Therefore, it is also interesting to observe the interaction abilities and emotions connected to it. David Rock, founder of the NeuroLeadership Institute, has developed the SCARF model that describes the social triggers of responses to the brain. They are helpful to understand interaction and behavior of human beings (Lloyd 2013; Rock 2008). SCARF stands for Status, Certainty, Autonomy, Relatedness and Fairness. These domains reveal essential discoveries of people's social interaction (Rock 2008):

- **Status** is about how a person is positioning itself within individual peer groups. Receiving an award within a group for example increases the personal evaluated position by the brain and thus sends out positive signals in the brain.
- **Certainty** reflects the brains capability of constantly evaluating situation and predicting future events. Very important to handle this trait of the brain well in an entrepreneurial situation, in which the future is uncertain and partly at risk
- **Autonomy** stands for the desire to control things. Using Stevenson's (1983) definition on entrepreneurship with *"the pursuit of opportunity beyond the resources that you currently control"* it becomes obvious, how important the autonomy aspect is for the entrepreneur. The interviewed entrepreneurs in the author's survey also mentioned the self-fulfillment as their most important motivation. A desire that requires a very high level of autonomy.
- **Relatedness** describes how humans evaluate their relations and safety position with others. As entrepreneurs act in this uncertain environment, knowing whom to rely upon becomes very important. Many entrepreneurs tend to trust few people and rather rely on themselves.
- **Fairness** is about the how fair situations and behavior between people are being perceived. A status that is unfair in the eyes of the entrepreneur could trigger the desire for change and free a certain energy, which is needed to overcome many entrepreneurial hurdles.

In the end, it must be a mixture of character, experience, talent, knowledge, connections and luck (compare 6.9.1 Hand of Fortune – described in a following section) combined with hard work that leads to success. This mixture and relation of elements is a basic principle of an ecosystem (compare chapter 3.5) and part of a holistic approach, in which the sum an interaction of elements as a whole is more than the independent summation. And many of the mentioned aspects are directly tied to the entrepreneur and thus form the center of this ecosystem. Eventually, the system setup and efforts spend to foster innovation and entrepreneurship should be so specific and focused that they help those, who have the abilities to become entrepreneurs, and not try to make everyone an entrepreneur. Many persons might not become entrepreneurs at all or be successful, because they lack the skills or have too many bad influences in their ecosystem. Those could be helped, because skills can be trained and parts of the ecosystem can be improved, or they could move to another country with a different ecosystem. Therefore, the Personal Circle of this ecosystem is about the entrepreneur's personal disposition to start a career as an entrepreneur, and his or her chances to further personally develop and to act successfully within the Entrepreneurial Ecosystem.

6.5   The Private Circle - Family, Friends, Neighbors, Coaches and Clubs

The next inner circle and closest area with a huge influence on the entrepreneur is the Private Circle. It consists of *family, friends, neighbors, the co-founders, social networks* as well as *coaches* and people at *clubs.* Studies have shown that people coming from an entrepreneurial background are more inclined to become entrepreneurs themselves (Janssen 2006, p.311); though this does not mean that they have to be necessarily more successful. But when people are exposed to entrepreneurs, co-workers or neighbors with an entrepreneurial attitude, they at least tend to acquire knowledge about entrepreneurial opportunities and become acquainted with customs that foster their own entrepreneurial process (Kacperczyk 2012, p.3). The influence of family, friends or neighbors for example, therefore needs to be considered broad and important. They coin certain aspirations within the entrepreneur. This can range from the desire to escape a certain environment to achieving the same standard then someone else has. Many role models originate also in this sphere. It starts in the early childhood on a subconscious level, when children observe their surroundings and the behavior of people (Shaffer 2009, p.19ff).

6.5.1   Neighbors

Neighbors can stand for an example of simple role models and owner of property that a child can observe – like the neighbor's house – and thus create an early desire for being like them or own items as they do (DeFilippis et al. 2013, p.126ff). This alone of course will not lead to more entrepreneurial spirit, because the material wealth mentioned above could come from professional person i.e. lawyers, doctors or business managers, with no entrepreneurial relation. However, if there is an entrepreneurial neighbor, children will recall the wealth connected with them. In addition, this desire for certain aspects like material wealth might influence a desire for an economic career. At the same time the base in terms of skill building, education and character strengthening for a general successful future is set already during childhood (Heckmann 2009). And this step towards economical thinking and the desire for success is an essential part for becoming an entrepreneur. Besides the material aspects, children might also be impressed by the stories of their neighbors and friends. If someone speaks highly of his father, a doctor for example, this could set positive seeds related to this profession (Susanne 2006). The same would than occur for a story about an entrepreneur. Another aspect is a potential motivation awakened in a child and especially in a teenager to do better than his parents and especially to do better than his friends or neighbors. This will form ambition and competitive attitude, which are also important aspects for an entrepreneur. Non "observing aspects" a neighbor could provide are advice or friendship, which will be covered underneath the section of coaches and friends.

6.5.2 Family

Social and economic research correlates the wealth of the parents to later success of the children in various fields (Mayer 2002, p.30). This could be a subconscious influence like described in the section on neighbors above. Aside from the subconscious influence, there are at least two other major aspects, where the Private Circle – especially from the family part – can influence entrepreneurship directly. The first is an open discussion and support for a potential entrepreneurial career. On the other hand, if for example the family continues to try to persuade or pressure a person into a career as a manager, it becomes counter productive to listen to those people, who care the most. This is especially the case, when this interference is wrapped into a context of advising to first collect experience as an employee, and then second, a few years after working, to start to rethink the self-employment plan.

Nevertheless, the financial support of the family can have a major impact on the entrepreneur – from the initial idea at the start and then throughout the years during the foundation and growth. This aspect was affirmed in literature research (Personal Financing 3.10.1), in the survey results (p.100) and can be confirmed by the author during his time at the business incubator CIE in Karlsruhe. Without the family funding and their backup, many startups would not have succeeded.

6.5.3 Friends

Friends also can play an important role at this initial stage, because they are the first level for social feedback and support outside the family. Proud parents often say positive, supportive things to their child, whereas friends can express a much more frank, critical and negative opinion. In addition, many friends can act as a mirror for the entrepreneur, who is striving for an entrepreneurial career. Friends may also become first time investors – as coined in the term "family & friends" financing round or be part of the founding team (Personal Financing 3.10.1)(Loayza 2013).

6.5.4 Clubs

Clubs are the extension of the mentioned aspects, when coming in contact with many more people forming teams. The team members, no matter whether it is in a game of chess, tennis or football, have an influence on the person (Martens 1996, p.51). Especially in the early childhood and at school, where character traits can still be partly formed (Dweck 2008, p.392). Observing team mates doing better may motivate to constantly improve, especially in sports (Bolter 2010, p.26). Training with a team helps a person to develop team player capabilities as well as leadership. Many clubs and the training within a team also have competitive and challenging components, whether in sport, in a chess tournament or in a math competition, thus in the end humans are very much influenced by the groups they interact with (Perner 1999). Likewise with the neighbors, the aspiring entrepreneur may also compare club mates and their personal background with his or her life, which could potentially lead to even greater aspirations.

6.5.5 Coaches

A coach, whether a trainer or team leader form the sports club or a teacher at school, can turn into a personal coach and mentor motivating and pushing the founder continuously forward in various disciplines. On the one hand, it is about improving skills, on the other hand, it is about training how to continuously motivate oneself and how to maintain a positive attitude (Bolter 2010, p.42ff). A good coach also catalyzes the club as an element of the Entrepreneurial Ecosystem, from training to learning social skills, and thereby demonstrating an interaction of elements within the ecosystem. An entrepreneur often listens closer and more to a friendly person not really related to him or her, especially if this mentor or coach knows what he or she is talking about.

All in all, the Private Circle Elements therefore stimulate and inspire the entrepreneur, while at the same time, the elements mentally and financially support, as well as train the entrepreneur's character traits and social skills.

6.6 The Educational Circle - From Kindergarten to Higher Education

The third area is the Educational Circle, ranging from **kindergarten, school,** to **higher education,** like college and university, as well as to **further education,** like seminars or training sessions. Also **internships** can be included to this cycle, because they have a huge educational component and usually take place during the education phase of a person. Two very strong motivating phases are probably the one in high school and the other one at the university or an apprenticeship. And it is not per se that the degree one can receive at a certain educational institution is of the highest importance. A number of successful entrepreneurs have for example never finished high school or college, but certainly have gained experience, input and inspiration during their time at high school

or college (Toren 2011). At these stages, a person's future is very much being build –
from skills to networking. Similar to the Private Circle, this domain has a subconscious
stimulus on potential founders, who are affected by the literature they have read, the
speeches they have listened to, the class mates they grew up with up with and the
teacher, who taught them. This circle has a direct connection with the inner Private
Circle, where class or sport mates become friends, where teachers become coaches or
where generally parts of this circle are seen as an integrated and important connection
to the close environment of the entrepreneur. The education system in the US and in
Germany, however, often concentrates too much on more memorization rather than
teaching kids and students critical thinking, problem solving techniques or collaboration
across networks (Wagner 2012). For innovation and entrepreneurship, it is essential to
show and to train the potential entrepreneurs and innovators to think outside of the
box, aside from learning the fundamental skills and basics.

An important influence could come from and often does arise through the contacts one
makes – especially in the higher education time (Pliz 2009, p.58). Not only do the
contacts subconsciously arise like described above, but also open doors to a world with
new impressions and new experience, as well as with direct opportunities to learn and
to work for. Education has also a third dimension – related to the name of the circle –
from which especially entrepreneurs can profit in a later stage, after they had chosen to
take the entrepreneurial path. Many skills entrepreneurs learn, for example at school or
in university, are later necessary to build up and run a company. Whereas
entrepreneurial education or courses have been in place and taught for a long time at
many American universities, German institutions are still on the upswing and need to
make up some ground (Global Entrepreneurship Monitor 2013). In this context, it is also
necessary to mention the student clubs and associations that have been built up at every
college or university in the US. Unfortunately, in Germany very few entrepreneurial
clubs exist, as well as very few events take place, in comparison to America that is full of
those activities, which actually cover all three dimensions mentioned above –
subconscious influence, contacts and skill training. Stanford lists 18 members to be part
of the Stanford Entrepreneurship Network (SEN 2014), MIT mentions to have 9
organizations in the entrepreneurship network and even 13 student organizations (MIT
2014). The University of Karlsruhe (nowadays called Karlsruhe Institute of Technology)
in comparison has three student club associated with entrepreneurship (KIT 2014).

Hopefully in time, Germany will follow the lead of America education in building and
forming the entrepreneurial character.

The career steps, taken by the peer group after graduation, have also a strong influence
on potential founders. Also when entrepreneurs get in touch with "opportunity costs"

and are de-motivated to build up a company, when friends tell them how much they earn as a starting salary. The challenge and daring to do something on their own right after graduation with no or little experience is a shot in the darkness. If too few peer group classmates dare to found a company, potential entrepreneurs can start doubting and asking themselves, why they have the requirements to start a venture, if their peers cannot or do not want to pursue this path of entrepreneurship. Today at Stanford University, there tends to be an opposite peer pressure towards entrepreneurship saying, "*if you haven't founded a company by the age of 20 you are a failure*" (Reuters 2011). With little or no entrepreneurial experience or a business track record, the decision, which track to take, is very hard to make. It requires external input from successful entrepreneurs, who, as sparring partners for the potential entrepreneurs, offer and help more than a comparison among their young peer group.

6.6.1   Kindergarten

One could argue that stimulation for entrepreneurship also starts best in kindergarten at the age of four, which many studies have identified as the time, in which a child's brain is reaching a perfect state for learning new things (Häring & Storbeck 2007, p.73). Children have their first interaction with other children, learn social behavior and can stimulate their abilities as well as build up first skills (Heckmann 2009). But since the time till the results for a potential influence can be really measured is lying at least 15-20 years in the future, it is very hard to analyze the influencing factors in kindergarten within the timeframe of the given thesis. Of course, everything that can be done to nurture the intelligence of children and to make them socially more competent, will help society generally – also in entrepreneurship – to attain to its full potential. Despite the influence on the development of a child in kindergarten, this thesis will concentrate on other elements in the Educational Circle, where systematic and structured learning starts, and therefore potential improvements are better applicable and measurable. Giving children the opportunity to learn and improve their skills is the foundation for important work in all types of careers.

6.6.2   School Years

What could be specifically done for entrepreneurship at the school stage, is to create a positive surrounding and to foster individuality, which encourages also among other things entrepreneurship and business. Schoolbooks should picture these topics from a positive and motivating side, influencing the youth subconsciously. These early emphasizes may later have a great effect for potential young entrepreneurs, as well as the people they deal with. All should be raised with a pro attitude and openness towards business and entrepreneurship. The author remembers that in his early high-school years, the entrepreneur was pictured in a French book as a fat bald-headed man with a cigar behind his desk, being harsh and unfriendly in the text to his employees. It did not hinder the author to become an entrepreneur, however the picture stuck with him, and unfortunately this negative image continues to be repeated by the media all to often. In social democratic countries, like Germany, TV shows, reports and movies often put businessmen and entrepreneurs in a negative light, in part to entertain the audience at the expense of the capitalist – or at least more neutrally speaking at the expense of the businessperson. In many cases, there is no distinction between bankers, managers and entrepreneurs, as group standing for greed and the exploitation of people (Huffington Post 2014). More on this topic will follow in the section on the influence of the media on entrepreneurship.

Even if these images do not hinder people to become entrepreneurs, they add extra hurdles in dealing with others that have been raised with a negative picture of individuals, who have become entrepreneurs.

6.6.3 Higher Education

The final years of high school and the beginning of higher education provide more options to directly teach skills and make contacts that open roads for entrepreneurship. The potential entrepreneurs often are motivated in class through their teachers, visiting managers, as well as entrepreneurs. In college or university they also meet potential co-founders and first employees. In addition, they can apply and test first entrepreneurial knowledge in seminars and competitions.

6.6.3.1 Basic Entrepreneurial and Business Skills

In many modern colleges or universities, aspiring founders learn the basic skills of management and entrepreneurship. Among these skills are the principles of how to calculate a business, how to evaluate a market, how to reach out to customers or how to organize and to run a project. Courses on project management, business case modeling and calculation or on sale initiatives could be taught. Many universities, which already offer these classes, are constantly improving the selection and quality of courses, while others are hiring new professors and teachers to install these basis courses.

6.6.3.2 Coding

One special skill nowadays strongly supports and increases the chances for being a member of startup and becoming an entrepreneur: coding. As many startups currently have an IT background or at least some components of it, this skill of coding is very valuable and offers a direct chance to start projects or join other young teams. This is actually encouraged by famous founders like Mark Zuckerberg of Facebook or Bill Gates of Microsoft (Code.org 2013). Although this group of entrepreneurs might have never needed further skills and stimulus to become entrepreneurs, there are thousands of other businesses, which are founded by entrepreneurs on a much slower and less innovative level, which requires all the help they can get.

6.6.3.3   Networking and Experience by Professionals

At this stage, potential entrepreneurs receive a lot of inspiration and insights from established entrepreneurs and managers. For this reason, many entrepreneurs are invited to institutions to hold lectures, in order to provide a strong practical component. These lectures cannot be overestimated. They are a great stimulus; they bring students together to reflect and to discuss the mentioned experience of the lecturer. Often as a result, an important peer group within the university or institution is created. These lectures also provide the potential entrepreneurs new contacts to the real market, because most speakers encourage further discussion. The Ivy League schools in the US and partly the few business schools in Germany have a wide range of interested people and alumni that are engaging themselves in and around the university. Investor networks, clubs and social gatherings offer insights and provide opportunities for the aspiring entrepreneur. The alumni networks, which students can become part of during their studies, are one of the key influencing factors in their career – whether it is for a management position or for becoming an entrepreneur. In the US, it is often said that persons return to university after 5 years of field experience, not only to do a Masters Degree, but also for the contacts that can be made. Once being in exclusive circles like Stanford or Harvard, many doors open for their alumni that for outsiders do not even exist. The alumni networks can be considered part of the social networks in the Entrepreneurial Ecosystem (compare Social Networks 6.7.10.)

6.6.3.4   Bachelor and Masters Degrees

The introduction of Bachelor and Masters Degrees was heavily discussed in Germany over the last years – and in part still is (Preuß, & Osel, 2012). Some arguments say that with the Bachelor degrees, the studies have become too structured and intensive, as well as filled up with courses, leaving little time for personal development, reflective or critical thinking. The students could not try out different things, while at the university, and often they failed "to live their dream", as Apple's Steve Jobs always recommended. However, what most pro and contra educational discussions lack is the opportunity perspective from a startup point of view. After the completed 2-4 year Bachelor degree, the students have the opportunity to try out a venture or work also for a startup at an earlier stage in life. After success or failure in work, the student can return to academia for a Masters Degree. During the time of the traditional European Diploma cycle of education, students had to study 4-6 years in a row, without the flexibility to start up a company, due to the fear of ending up with no academic title. It was an all or nothing situation, thus making it very complicated to experiment and gather experience in the market, when working for a startup. Once finished with university and a Diploma degree, the students have spent so much time at school that many are tired of a few more years on a low and insecure income. In addition, they cannot go back to university after a failed startup, but rather have to apply for a job, which leads to doubts, how the labor market evaluates their time in a startup and especially, when there had been a failure. No doubt, the Bachelor Degree helps to overcome those reservations and timing problems. In case some universities return to the traditional Diploma studies, they must make sure, to introduce an intermediate degree, which is internationally accepted. This will most likely lead back to the Bachelor Degree.

6.6.4   Internships

Lectures from the mentioned external experts, specialists or alumni bring the students closer to the business market. However, internships, student jobs and projects are a much more intense step for inhaling "market air" and are a first source of practical experience. Internships are also a form of education in real market situations. In part, it is learning by doing as well as being taught by colleagues and employees at the company. This certainly is a crucial step for any businessperson, especially for entrepreneurs, who often in the past need to work in all different fields of a company. Often, the lack of employees forced persons to take on additional special functions. Founding a company, finding customers, working with others etc. requires to make use and improve abilities and skills at a level, one can never learn in theory at school. First market knowledge is gained at this stage, and contacts to potential partners or customers can be initiated at those internships as well.

6.6.5   Further Education

Further education is the in the author's opinion an underestimated and wrongly approached aspect. Since the potential entrepreneurs choose this form of education usually themselves on top, one can expect much more attention by the entrepreneur to the content taught as well as to the people the entrepreneur gets to know. In this context, further education ranges from seminars, to educational programs and to executive master programs. Many Ivy League schools consider their general master programs as the essential part of a person's career (leaving it open, whether a master program only seeks to provide to higher education or further education in the entrepreneurial circle). But even more limited programs like coaching and training sessions or weekend seminars, which entrepreneurs in Germany have to attend, when they receive a public scholarship for example, are useful, informative and provide networking. In both cases, it is important to receive a stimulus and advice from people, who have a profound knowledge in their field.

When striving for a management position, certainly some experienced managers must teach some of these courses. This turns out to be often the case in the US, at the top business schools and elite institutions. When it comes to teaching entrepreneurship, US schools often invite successful entrepreneurs from their alumni pool. In Germany on the other hand, the coaching and training for entrepreneurs occurs mostly by persons with no entrepreneurial experience. Worst case, this is done by a person with neither business nor entrepreneurial experience. But even an experienced corporate businessperson does not know, how it feels to approach a customer with no references, how to build up an unknown brand or how to operate with hardly any financial means. These corporate managers have never used their own money to pay the startup and an employee or have not turned an idea into practice by themselves. Everyone would agree, for example, that a pilot student should be taught how to fly from a pilot, but when it comes to entrepreneurship, it seems that a theoretician is as welcome to teach others young founders as a practitioner. All the teaching and training from local, regional or federal institutions should therefore be reassessed. When it comes to the design of entrepreneurial concepts and the distribution of money from official sources for entrepreneurial programs, coaching and other initiatives, it is also necessary to have people with the right qualification in the driver seat. Besides having less successful entrepreneurs that could fulfill such practical tasks in Germany, it is also the different culture that leads America to have more successful entrepreneurs and business people, involving themselves publically after reaching their personal goals and living their dream. Many incubators, at universities in the US, are initiated or run by entrepreneurs; and even big public offices like the major position in New York was run by Michael Bloomberg, one of America's most successful entrepreneurs.

6.7   Public & Business Circle – From Culture, to Economy and Geography

In the Public & Business Circle the entrepreneurs finally enter the stage of action. It is the widest and most complex circle including culture, government, history, language, geography, media & information, the finance world and the economy & business world. The Finance and Economy & Business World Elements are in this context of the Entrepreneurial Ecosystem considered ecosystems of their own, with many elements and rules from their ecosystem affecting the Entrepreneurial Ecosystem. The Economy & Business World Element for example includes classic aspects like market maturity, economic growth, infrastructure or concrete business aspects like supplier and customer situation, as well as business tools and methods. Of course, it also shares common elements like culture or government with the Entrepreneurial Ecosystem. In general, some other elements such as a government or higher education have also ecosystems of their own. However, many of their elements and rules do not affect the Entrepreneurial Ecosystem, and are therefore considered to be only "simple" elements and not elements, which are also an ecosystem of their own in the context to this Entrepreneurial Ecosystem.

6.7.1 Finance World

The Finance World Element of the Entrepreneurial Ecosystem plays a significant role for venture-backed startups. It plays the key role in supporting the startup activity. If not financed through direct sales, entrepreneurs will always need financial support from banks, investors or their family and friends. The importance of this element cannot be overestimated and has shown a clear difference between the US and German support for its entrepreneurs.

A strong venture capital and business angel base increases the success factors of startups enormously, because they can provide the decisive financial assets for technology development, market entry and growth. Especially with the huge difference between the US and Germany in terms of funding capacity, it is a tangible and a quantifiable topic.

Particularly the early stage financing, as explained in Chapter 3.10, is one of the crucial aspects for startups that require initial funding. Most Internet startups as well as all companies that require a high volume of research and development belong to this category. Startups in a service or trade environment or with a quickly to implement product or solution, often have the chance to create direct sales. They can either live partly or fully from these revenues or have an option to receive loans from banks or their family.

Nevertheless, almost all companies have hard times receiving money from foreigners or strangers, where their family, local banks or native investors cannot be counted on. Most young companies have low or now revenues, credibility and proof of concept, making it very difficult for outsiders, to understand and to believe in the company and team. As explained earlier in the buyer's and seller's situation, these difficulties of receiving money becomes even more difficult, the fewer investors are available on the market. At the start of a venture, many founders are often caught in the dilemma of having no real proof of the functioning or need for their business, which is essential to convince investors to fund them, while on the other hand, the market situation often does not urge investors to invest, before another investor would do the same (Überla 2011, Romans 2013). This will lead to less investment deals, a very high cherry picking factor and lower evaluations. This situation can mainly be observed in Germany (compare 5.2.2).

Another reason for the lower evaluations in Germany and Europe are also due to the lower exit channels. Since chances of exits in the upper 2-digit or even 3-digit millions of EUR are rather low, investors see no reason to do investments on high evaluations, because the return on investment with a potential exit is not worth the risk (Manger 2011). This leads in the following to further problems. Not only do fresh startups have problems finding money, it also costs a lot of time to do the funding. A time loss, which affects companies in later stages, when follow-up investment is needed, and thus eventually hinders them in becoming fast global players. Especially in the Internet sector, this is a crucial aspect, where speed is a necessity for global leadership.

For the author, the lack of a wide funding opportunities lead to another even more severe problem. When investors start cherry picking the potential best startups at a certain step as well as those, with the best track record, many startups do not get a funding that could have changed their business with a twist and the provided money. And since it is very hard, not to say impossible, to foresee the business future, the best performing and blockbuster companies of the future might not get funded at all. A situation German politicians are constantly discussing and for which a solution has not yet been found (Deutscher Bundestag - Unterausschuss Neue Medien 2011). However, part of the answer can be found through the use of statistics. The pure number of startups that receive funding in the US is so much higher that out of promising 1,000 startups for example, 100 might become high performers from which 10 could become "game changers" and global market leaders. At the same time, Germany, with 1/50 of the amount of funded startups, would see in the drawn comparison 20 promising startups, with 2 high performers and 0.2 blockbusters – which actual means none.

An economy or nation needs as many startups as possible in the first place, in order to have more options to pick the best performing companies later. It can then let the individual companies proceed and wait, to see how they and their customers develop. Once this positive high-volume funding cycle has been established, the performance or success rate of picking and pushing the right startups will also improve. This would be due to more investors with special expertise that can help the startups to be better and faster. The networks will be more connected, the openness for venture capital and startups will increase, and with more successful startups and exits, there will be more successful individuals and founders that can become business angels and investors themselves.

The Financing World Element is therefore one of the essential parts of the Entrepreneurial Ecosystem, with a high influence on startups and innovation.

6.7.2   Culture

Many papers and sources on entrepreneurship mention that culture has one of the greatest influences on an entrepreneur (Global Entrepreneurship Monitor 2013). If society, from the customer to the businessperson, is not fond of or open for startups and young entrepreneurs, a chain reaction can be expected that leads to difficult conditions for new ventures, less investors and less support. One can almost speak of a vicious circle, when few founders find only few investors, insufficient suppliers and a small group of customers, leading to fewer success cases, and thus few new investors for the new founders. In this situation, it is difficult to build up a widely spread positive image of entrepreneurship, which would be a necessary part to break the vicious circle. It is quite obvious that a pro-entrepreneurial culture can influence many elements and speed up the necessary measures. A pro-entrepreneurial culture will be willing to spend more money for necessary programs, it will lead to more open doors for young entrepreneurs and create more positive stimulus for startup initiatives. Culture itself is also influenced or based upon other elements of the Entrepreneurial Ecosystem such as history or geography.

6.7.3   History and its Connection to Geography and Culture

Chapter 5.4 on the comparison of USA and Germany has shown, how history and geography are connected and how they influence entrepreneurship. The pattern of opportunities through geographical, cultural and historical influences can nowadays also be seen in the BRIC countries, where their potential is even higher, due to the mere size of their populations. And countries with a huge population like Indonesia could follow the BRIC, indicating there are more to come. It is important to examine the maturity of an Entrepreneurial Ecosystem, in order to understand, in which phase the nation is. The following aspects are derived from the comparison between the United States and Germany:

- The mixture and origin of the people (Have they been fleeing? Do they have anything to lose? What assets, talents and culture do they bring into the country?
- Culture (What is their philosophical, religious and cultural background? What kind of natural characteristics do they have? Is there a pro-entrepreneurial culture? What is the general attitude?)
- The development in the country (Is there an upswing, downswing or wealth-maintaining situation? Is there an established infrastructure or an open territory? Is collaboration among the inhabitants needed? How mature are the business segments?)
- Isolation or continuous mixed of influencing factors (What impact did war have upon the nation? Are there new immigrants? What is the cultural mix? Is there a unified system, one language or one currency? Who are the allies?)

- The global and regional ambition (Is the country fighting for global or regional territory? Is the country involved in international conflicts?)
- Geographical position (Is the climate at certain locations forcing the country to innovate? Are the environment and territory issues influencing the business side?)

It must be remembered that in Germany the situation after a few decades has gone back. After the successful rebuilding of the country and after the new growth of wealth, the entrepreneurial aspect moved backwards, overrun by the desire for security of the acquired new wealth. This change caused the upswing to a more "wealth maintaining" situation. Nevertheless, this offers improvement potential from an entrepreneurial perspective, when seeing an ongoing drive for entrepreneurship in the US, despite a high level of wealth.

It is almost seen as a given fact that Americans have a culture of entrepreneurship, which is greater and more advanced than that of Germany. This difference is contradictory to the German *Wirtschaftswunder* after World War II (BMWi n.d.). During the rise of the German economy, many SMEs occurred and became strong international market players. It required a lot of entrepreneurship in those days, since structures and products had to be rebuilt and businesses had to started from the scratch. What caused Germany to lose this kind of entrepreneurial attitude? In a way, it has not! Germany is still known for its great *Mittelstand* and its powerful SMEs that are often world market leaders (Wirtschaftswoche 2013). And many of these had been founded right after the war or grew significantly after it.

However, one can observe a certain affinity to security, high quality standards and a healthy modest growth attitude in the German business culture. This culture has been part of the German culture and had been nurtured already during the time of Bismarck and Prussia. As explained in chapter 5.4 on German and US history, in those days the German educational system had the interest in forming and in educating good loyal workers and officials, which can also be seen in the German educational system till today (Precht 2013). Having lost almost everything due to World War II, Germans, like Japanese, rose to the occasion and restarted from almost zero, leaving behind also many traditional, limiting values. At that time, it was the ability to change, to restart and work hard with their backs against the walls that helped the Germans. A good example of the need is houses of the early 1950. Housing was desperately needed, while time and money were very short. Thus, many houses were built in a way that many Germans today consider them a "constructional sin".

Interestingly enough, the same can be observed in China today, which is critically viewed by many builders in Germany, since the buildings are being built too rapidly with low standards and poor quality (Die Welt 2012). It must be remembered that for a nation on the rise, often time and money for high quality growth are lacking. Once Germany had regained a certain status and standard of living, the old patterns of security and quality swung back. Probably an attitude that can be found with many humans, because once there is something to lose, people try to maintain it. Whereas having nothing to lose, forces more flexibility and a greater need for positive change. Entrepreneurship stands for change and trying new things under uncertain, difficult conditions and questionable outcomes. This is where Americans indeed have a more flexible attitude. It seems, most Americans believe they are able to change their situation at any time and create a better future for themselves.

This is a perfect base for entrepreneurship. Despite the numerous definitions of what an entrepreneur needs to do to be called, many Americans simply call themselves entrepreneurs, neglecting those prescribed definitions. It is their attitude that makes the difference. And this sets the ground for a positive entrepreneurial culture. On the one hand, many consider themselves as entrepreneurs, while on the other hand they also believe that everyone can make a change and difference, thus there is generally a stronger trust in the success chances of an entrepreneur. Since a nation is based upon a majority of people with a certain mindset due to its history (compare chapter 5.4), it may explain to an extent these differences between Germans and Americans. An interesting fact is hereby that about 25% of US Americans is of German and Dutch descent (Business Insider 2013). Without doubt, Germany has lost a considerable number of its entrepreneurial population through emigration to the US, and thus needs to invite many entrepreneurs from all over the world to build their future companies in Germany.

6.7.4 Geography

Geography itself can also play a significant role. Founders, in small countries like Israel or Singapore for example, have to prepare their startup for an early internationalization, due to the small domestic market. On the other hand, startups in isolated bigger countries or a continent like Australia have to look for other geographic regions, where they can sell their products too or buy their resources from, in order to grow. Running an E-Commerce business with an international audience can be, for example, very complicated for an Australian startup, due the distances to other countries and the high shipping costs.

Countries, surrounded by many other nations, have the advantage of receiving input and talent from various sources, yet they have to manage the complexity of different languages, laws and currencies. European countries are good examples for that.

But also, a country's involvement in political international conflicts and interests due to geographical situations, have significant influences upon startups and business. Israel is an example of this situation, with wars and conflicts in their region, causing multiple to government spending especially in new high-tech and military equipment (Ministry of Economy State of Israel 2014). Coming from a different perspective, but also with a high influence, the US has plays a certain role and, like all countries, has a special interest in the defense of its borders. The government spending for pre-commercial investments at around 200bn USD in 2011, also helped many young companies to develop and sell high-tech and security relevant technology to the US government (European Commission DG Enterprise & Industry & ECORYS 2011, p.7). Part of this, is certainly linked to its history and politics, but parts are also due to its geographic position that leads to domestic and international investments. The expenses range from space technology, to military field equipment or to new genetical and other agricultural products.

Today, many founders also discuss, whether the weather especially and the environment generally has a positive or negative influence on their performance and thus on their startup. Silicon Valley could lead in this category, offering long summers and many sunny days. However, some studies for example from Harvard Business School almost propose the opposite, saying that rainy days are better for productivity (Lee et al. 2012). Yet, they have not examined the motivation aspect and not the perspective of an entrepreneur. Since there are many other influencing factors that can make the Entrepreneurial Ecosystem strong, as we can see at Stanford University, the weather phenomenon requires plenty of more research, before being able to give an adequate answer on its influence.

6.7.5   The Influence of the Media

Culture plays a major role for the Entrepreneurial Ecosystem. Even though this culture of established economies has been developed over many decades or even centuries, it is constantly influenced by ongoing developments. The key messenger of these developments is the media from newspapers, to TV or Internet content.

6.7.5.1   Media Coverage

The mass media without doubt influences the public opinion, and thus, in the long run, the culture. It impacts and also influences the business world, by helping young companies to have a highly visible and positive image – Spiegelonline, Wirtschaftswoche or Handelsblatt in Germany for example are frequently reporting on startups and their function as media partners in national and international competitions (Spiegelonline 2014b; Handelsblatt 2014b; Wirtschaftswoche 2014). Germany pales against the US when comparing the media coverage of startups. When searching the term "startups" on the website of the NY Times, one receives 55,398 all time results (till the end of 2014), whereas the German FAZ website delivers only 479 search results for that term (New York Times 2014; FAZ 2014). Each of these NY Times and FAZ press articles brought more publicity to the startup. This kind of publicity is important for the selling of products or building up business relations, as a young inexperienced company. The influence also extends to the finance world, since the awareness of startups can encourage to more people to engage themselves as investors.

6.7.5.2 Transporting a Negative Message about Entrepreneurship

Journalists often state that they mainly transport opinions of others, however, as seen over the last years, it is more than that. Journalist indeed can fuel and steer a topic or discussion into a certain direction, and they can influence people through their reports. Their responsibility for certain developments is, therefore, much higher than proclaimed (taz 2014; Noelle-Neumann 1996). When it comes to entrepreneurship, journalist write negative comments on the developments in the economy, with some provocative or not understandable behavior of business people, their actions and salaries. These remarks can and often do also reflect poorly on entrepreneurs. Managers, bankers and entrepreneurs are often thrown or mixed up together.

The negative theme is that a few people have too much power, earn too much money in comparison to the average employee, and that society even needs to pay for their actions, if they steer the company into the abyss – something of which already Milton Friedman has warned against in the 1960s (Bloomberg 2012; Friedmann 1962). Despite the fact that there are also only money oriented, selfish and reckless entrepreneurs, most incidents that led to a public outcry or financial problems burdened upon governments, were caused by managers and bankers, who do not own or have founded the company they are working for. Therefore, it is very dangerous for entrepreneurship to be thrown into the same boat. Entrepreneurship is based upon an individual that is trying something new out, for which he or she will take full responsibility as well as collect all the praise and merits. The nature of these actions and results is certainly everything but average. One will hardly find an unsuccessful entrepreneur, who leaves his company with a "Golden Handshake", in spite of the fact that having just led the company to suffer a major loss. Unfortunately, the media in Germany, not only mentions entrepreneurs in the same context with managers and bankers, it also picks out the worst and dubious entrepreneurs to underline their message of greedy reckless entrepreneurs and managers – such as the former owner and CEO from AWD Carsten Maschmeyer (Spiegelonline 2013).

6.7.5.3   TV and Movies

If one counts TV and movie entertainment also to play an important role in the media, it becomes necessary to observe, how business and entrepreneurs are dealt with and used in their stories. Here, US and German entertainment content on this subject does not vary much. Despite the fact that many movies broadcasted in Germany have their origin in the US, also the German entertainment media uses negative clichés, such as rude entrepreneurs laying-off employees while living in luxury. Rarely stories are told about good business owners being concerned about the welfare of their company, the employees and the customers at the same time. Many movies use the same pattern of entertainment and report as journalists, who only see the dark side of the business world and its owners.

Unfortunately, banking, Hedge Fund or Private Equity business are the subjects used, in which the managers, in order to achieve their goals, focus purely on money and thereby ruin other people's lives. Examples of this are movies like Wall Street, Margin Call, The International or even Pretty Women. Although in most movies the main actor, ultimately, either gets punished or becomes enlightened for a better cause, the negative image stays that business people are ruling and ruining our economy. They are mainly ruthless and not acting in the interest of society. An American audience often tends to be more flexible in their interpretation of this, and of being able to live with the contradiction of loving and hating this rich group of individuals.

6.7.5.4   Different Reactions in Germany and the US

Americans are faced with contradictions every day and seem to like it (Gelfert 2011; The Tech 2003; The Washington Times 2013). For example America is proud of its freedom of religion, while the president of the United States often mentions "God bless America". Or even on a US money bill one will find the quote "in God we trust". Americans love capitalism, they believe every one is responsible for their own fate and consider Europe and Germany as a form of socialism with many left leaning welfare states. At the same time, the United States has the highest amount of donations and foundations for social projects in the world (Charities Aid Foundation (CAF) 2013; The American n.d.). It is one of the most prude countries, while at the same time hosting in California the biggest porn industry in the world. Americans can overlook and deal with this, but hardly the same can be said for Germans. That is why picturing business activities in a negative light, has a rather negative effect upon the German society, and makes it more difficult to ignite more sympathy and positive understanding in general for business and entrepreneurship.

The extravagant entertainment image of capitalists certainly does not improve the picture of entrepreneurs in public, thus causes difficulties for their personal and business life. Especially in Germany, as elsewhere, the envy culture will thus be nurtured, shedding a negative light on everyone earning highly above the average worker. Entrepreneurs are indeed very social beings that are affected by public opinion, because it is part of their reward to get positive feedback. A positive sign can, however, be observed in several on- and offline newspapers that regularly report on startups, their struggles of finding investment or their outlook in terms of development – such as Spiegelonline, Zeit.de, FAZ, Wirtschaftswoche or Handelsblatt. This upward positive movement needs to be nurtured and spread throughout the media landscape, and thereby making startups also an important part of the daily business life.

6.7.6   Economy & Business World

Being an ecosystem of its own and making up the core of each business, the Economy & Business World Element offers many contact points and influencing factors for the entrepreneur. However, not all of them are crucial for the founding process of have a specific entrepreneurial context. This element represents also the gateway, where a startup becomes an established company. An important influencing factor can be observed when looking the so-called BRIC countries that have huge domestic markets, underdeveloped structures, low standard of living, and thus offer plenty of growth potential.

In the more developed Western countries, some business conducts and available tools can also have a negative influence. For example, static business plan structures and approaches slow teams down or even discourage them. The lack of various scenario techniques among many young entrepreneurs and the lack of financing causes "small thinking and acting" in Germany and elsewhere, and thus hinders global thinking and leadership. The base can be changed in the Educational Circle, the application and acceptance of it, however, is in the market, and thus part of this element.

Infrastructure is also an important factor of this element, which can help and hinder the development of a startup. Without Internet, for example, one could not grow an E-Commerce company. However, emerging countries like India with a more or less bad infrastructure still show high levels of entrepreneurship – as Germany did after its destruction in World War II. Thus, infrastructure and its influence on entrepreneurship is a very complex matter.

The following table will show some of the influencing factors for entrepreneurship in the Economy & Business World Element (Ortmans 2009; Woolley 2011; Fairlie 2011; McFarlane 2013, p.59; GTAI 2014; Dowling 2002):

Aspect	Influence on entrepreneurship
Business tools & methods	- Global understanding of Business Plans (e.g. leading to high work load when writing it) - Commonly usage of business cases and scenario planning (e.g. a must for many businesses and especially when seeking funding) - Huge variety of IT tools at low prices also available for startups and small companies (CRM, Mail, project management, ERP etc.) - Accounting and financial standards (e.g. make global business possible yet increase also complexity for small companies: IAS, EBIT, ROI, marginal income, profit & loss) - Availability of management techniques & tools (e.g. SWOT, Porters 5 Forces, 4P Model, value chain, business model canvas, risk analysis, company evaluation schemes)
Customer, supplier & partners	- Business relation and conduct between business players (e.g. procurement policy of corporations hindering startups to offer their services, expensive standards like SAP interfaces or PMI certification, credibility of startups [compare cultural aspects] financial track record and required levels of turnover to participate in certain businesses etc.)
Economic growth	- Increase of purchasing power - Creation of new business opportunities - Creation of a pro and positive mood - Hyper growth leads to high evaluations and investment flows - In emerging markets less and low standards offer opportunities for the founding process, however could also hinder investment due to higher risks
Established/ wealthy economy	- Luxury and high-quality products/ services on a large scale - After sales, recycling, repair and maintenance - Increase of individualization creating new desires - High purchasing power - Market satisfaction in many areas - High and established competition

	- High costs of personnel, war for talent - Higher security of assets, technology and data as well as personal safety - Less corruption - Established judicial system - Better infrastructure - Better educated personnel
Financial system	- Provision of financials through banks or risk capital (compare Finance World Element)
Infrastructure	- Basic supply (e.g. stable energy, water or Internet is a base for businesses) - Transportation (e.g. stable, safe and affordable nationwide transportation system for persons and goods is crucial for most business activities) - Functional services (e.g. functional postal system necessary for online marketplaces; stable bank system necessary for commerce business) - New business opportunities (e.g. fast and ubiquitous internet or 4G mobile makes new services like multimedia content, real time cloud services or location based services (LBS) possible) - Public services & safety (e.g. Police, hospitals and public authorities guarantee safety, fair competition and protect the business and its employees)

Table 35: Influencing Aspects from the Economy & Business World Element

6.7.6.1   Business Plans & Business Cases

In many business arenas, whether it is at business plan competitions, at university or at a consulting agency for entrepreneurs, the business plan is thought to separate the founder form the wishful dreamers. A solid plan opens the entrepreneurs' eyes for all prospective challenges and unfulfilled tasks, which await them on the way of their companies' realization and throughout the growth. It is required in many occasions, and thus is an essential part of the Economy & Business element. The current models are very intense and complex and often very hindering and demotivating.

A new setup, understanding and timing is, therefore, required, to help the entrepreneur develop and plan the business properly. Looking at the investor's need for a profound Excel model, one can already see some answers: From the revenues deriving from sales-quantities and prices, to the necessary spending for personnel, marketing and infrastructure. In order to be more profound, the entrepreneur tackles virtually each topic, which a business plan and case is looking for. Even the reaction of potential competitors will be found in a good Excel model, by changing of prices, increased marketing budget or decreasing sales units.

6.7.6.2 Tools & Business Methods

There are several tools and techniques that entrepreneurs need to use throughout their founding and growing phase. They are part of the Economy & Business Element/Ecosystem. The one with the strongest influence is the **business plan** (compare 3.13), yet other skill-based methods like **market analysis, scenario planning, business case & cost calculations, presenting techniques** or **project management** also play an important role. Most of these aspects actually find a place in a business plan. For the founder, it is important to be aware of the various tools and the timing, in order to apply them, when they are needed and integrate them well in their business planning.

Example:

> A founding team tries to evaluate, whether their business idea has potential and what kind of concept is needed to implement it. Their idea is a cell phone cover made out of recycled material. In order to do an evaluation, the team consciously or unconsciously applies a form of **business case calculation**. Usually, it starts with a micro calculation of a person or group being interested in the product and willing to spend an amount of X-Euros. The team decides to charge 20€ for the cover. The customer is an environment-conscious person, who likes to protect mainly his or her cell phone from scratches and damages. The second step will be to transfer this small entity to a larger group in a certain area either on a local or national level. Since the team is located in Germany, they decide to focus on the business area of smartphones there, because they believe that for such valuable phones, the people in Germany, will pay extra for a protecting cover. This step requires a short **market analysis**. From the 15m smartphone-users in Germany, they think 80% are interested in such a protection and 50% of them are very environmental oriented. The team has taken these percentage numbers from the number of Germans constantly buying organic products, which lies according to several sources between 30-50%. This will eventually lead to a market value of 120m EUR for their particular recyclable cellphone cover. Further steps could be

adjustments in terms of extension or change of the group, increase of the price per unit or adding further services. The next step would be an estimate, how often customers will use this product. As the average cell phone contract for high valuable phones in Germany is 2 years, the team applies the same cycle for the cover purchases. It is often heard that startups need only 1% of this market value to be successful – and 1% always seem to be feasible. In this case, it would mean the team could make 0.6m Euros sales per year, if the 1%, of the two- year cycle buyers they reach for, buys every second year a new cover.

As soon as a detailed *cost calculation* is added to the *business case*, more elements and assumptions need to be made, in order to fine-tune the business case. This case method approach will be adjusted and carried out in detail throughout the development of the startup. But when it comes to the point questioning the first assumptions, further tools and techniques are required.

A more profound *market analysis* in this case must follow. The team would, therefore, need to scrutinize, how many smartphone users really exist, how that user number will develop and how much a person would be willing to pay for the protective telephone cases. The market analysis also helps to attain more certainty about the actual costs. The costs for marketing, personnel or IT, for example, can be evaluated through a good market research. Calculations for procurement costs could also be done through the market analysis.

Major errors in the case calculation or market analysis have a high possibility of causing the startup to fail – due to lacking of investment or unsatisfied customers as well as founders. Investors will quickly realize the wrong assumptions or missing information. Customers might find the pricing of the product uninteresting or do not see the advantage to competitors' products. And even within the team, false assumptions can demotivate the members, due to unexpected developments.

The closer the team approaches a market entry, the more important a proper *project management* will become. On the one hand, the complexity will increase with more tasks and limited resources. On the other hand, teams have to plan their operations, in order to convince potential stakeholders like investors that they are capable of implementing the job. At this point, the proper application and use of the skills to provide a *business plan* with potential *scenarios* are essential. Although, as prior discussed, business plans can change or adapt, teams having a proper business plan might not be more successful, as teams without one. Furthermore, convincing investors to invest in the team, always requires a collection and presentation of essential information. This collection, in many parts, aims for data from the business expectations, as well as for impressions on the team. Once the business concept is ready to be presented, **presentation skills & tools** are required. Many startups do not have a group of experts, and especially software and traditional engineers are having problems to acquire business skills required to run a company. Aside from knowledge on how to structure a presentation, tools like the classical PowerPoint or a newer web based service called Prezi are necessary, to technically implement them. When it comes to the holding of the actual presentation, not only skills of guiding a person or a group through a presentation, but also talent in demonstrating it and speaking in front of an audience play a role.

In addition, it is also very important to be able to outsource and delegate tasks. Prof. Faitin (2008) is in this context argues that entrepreneurs do not need to possess all the relevant business knowledge, because most tasks can be outsourced. The founder should concentrate on the core aspects. However, the author only agrees in part with Faitin, because being able to outsource things requires also the knowledge of their existence, their necessity and how to control the problems that arise. The question is also, whether this aspect is more character trait or a trainable skill.

6.7.6.3 Markets & Economic Situations

The present, actual status and development of each individual market, as well as the state of the economy have a major influence on the success of a startup. Thus, they are part of the Economy & Business Element/ Ecosystem. A strong influence can be seen from the investment side, when an economic downswing like the "2000 .com bubble" was bursting. Venture capital money decreased immensely. On the one hand, because many individuals, as well as funds had lost substantial amounts of money, on the other hand, because the crisis was to a great extend about money going into startups that were failing (Hendrik Brandis & Whitmire 2011; Investopedia 2009). In the following years, it was much harder for startups to receive initial funding or VC money for further growth, even in the US. The German market interestingly enough took much longer to recover from that time, because VC was at its beginning and nearly crushed by crash in 2000 (Venture Capital Magazine 2013). Later, Block and Sandner (2009) found out through analyzing US venture investments of the that there was indeed a 20% decrease of VC money, due to the financial crisis of 2007/2008.

According to their research, this decrease affected only startups in later stages looking for further funding. Younger startups simply needed to postpone their initial funding, till the markets turmoil had relaxed. Since such large economic downturns ("bursts") happen quite frequently (e.g. 1992 Japan crisis, 2000 .com crisis, 2007 financial crisis), startups have to be aware of economic ups and downs, and either adjust their timing or live with the consequences. In terms of timing, VC Tim Draper optimistically believes that the upcoming years are great for building up startups, and he thinks this could last at least until 2017 (Romans 2013, p.6).

Another strong influence upon startups is the economic situation itself. Emerging countries like the BRICS (some sources include nowadays also South Africa to the BRIC) offer zillions of opportunities for new companies, as well as investment funds interested in investing early at high evaluations in hope for great returns on invest (Sawhney 2011). Despite political and social risks, as well as insecurity in these markets, the pure size and developing potential compensates for the risks and lets the money flow easier into these markets. On the one hand for not established businesses, which have shown their proof of concept in the *Western* countries; on the other hand for totally new approaches and businesses, which are required because of the regional specialties, the mere size of the BRIC countries and the need for catching up with the far more developed countries.

In addition, it is not only about the investment provided, but also directly about the increasing purchasing power, which leads to more demand in many fields. This increase in demand gives startups chances in various industries and fields. A situation, which can be compared to Germany after World War II, when almost everything was needed and had to be rebuilt. This time after the war was good for entrepreneurs and the creation and growth of many companies. This led to what is known as the *German Wirtschaftswunder* (Encyclopædia Britannica n.d.).

Today, mature markets like Germany or the US do not offer anymore this high opportunity-risk relation in their own country, and thus it functions differently. The advantages of these mature markets are large customer groups with a high purchasing power and a very advanced business infrastructure. On the one hand, startups might need to look harder for niches or make their products much better, in order to become more competitive. On the other hand, they can build upon established structures and processes – from the sales channel to business cooperation. Nevertheless, the emerging countries also play a role for startups in the mature markets, when it comes to expanding their business abroad. America maintains its advantage, because of its leading role in the business world. A successful US venture can be easier transferred onto other markets, because it has proven itself in the biggest market, is operating in English and probably already has started to build up ties to several other countries, which have traditionally close relations to the US market. These aspects not only apply to expanding to emerging markets, but certainly also to growth in other mature markets. German startups have the disadvantage of a smaller home market, which does not give them enough financial relevance. They have to operate on the other European markets, in order to achieve that necessary international relevance. Different languages, laws and cultural differences make this step for EU startups more difficult, than for a startup from California that also wants to do business on the East Coast.

6.7.7 Language

Another factor with a huge influence is the language. Teams that operate in English speaking countries are automatically closer and more open for international growth, without extra cost and effort. First, because English is the world business language; second it is spoken by 1.75 billion people at a useful level (Neeley 2012); third, there are 500m native and second language speakers in 105 countries (Nationsonline.org 2014); and fourth, because countries, where English is spoken as a first language by the majority of the population, account for 25% of the global GDP with only 6% of the global population (Nationmaster 2014). This can be observed not only in all US firms, but also with foreign firms coming from small countries that force the teams to operate in the English language such a Singapore, Israel or the Baltic countries. Their home markets are so small that they plan to internationalize from a very early stage on (usistf.org 2014).

But even large countries like India have stressed the important business communication in English, not only because they are a former British colony, but also they want to position themselves with the global English business language as an international company that is open for international markets and investors everywhere (Sawhney 2011). However, if the domestic market is large and strong enough, like China or Brazil, the language becomes less important, at least for the early growth and establishing years. Companies can make millions in sales without encountering other languages and nations. With globalization this is nevertheless more and more changing. Even in an "early stage company" language can play an important role for international contacts, suppliers or customer communication. And the Internet opens international sales channels, whether it is for products being shipped or for IT services being digitally transferred (Roland Berger 2014). And all these channels operate in one language, English.

6.7.8 Government

The Government Element sets the political and legal framework for all entrepreneurial activity. It can hinder business dynamics but has also the chance to stimulate it. Although it is not valued high by many entrepreneurs and investors, there are quite a few activities that – if done wisely – can foster entrepreneurship, and will therefore need to be further analyzed in the following sections.

Eventually, the finance world can have the key role to fuel the startup activity; if not financed through direct sales, entrepreneurs will always need financial support from banks, investors or their family and friends. If these financing instruments do not function properly, the government can try to fill in the gap. The importance of the Government Element in general cannot be overestimated and has shown a clear difference between the situation in the USA and Germany on the financial support side by the government.

Government programs can try to bridge malfunctions of the Entrepreneurial Ecosystem and may replace missing elements. Due to the complexity of this ecosystem and missing comprehensive understanding, these programs are mainly focused on financial support. In Germany, there are scholarship programs like EXIST "Gründerstipendium" from the Ministry of Economics; programs for subsidized loans from the KfW or investment matching programs for investors from both of these institutions. They will be described in the following subsection. The largest effort currently seen in Germany is the government fund HTGF (High-tech Gründer Fonds), which in its second edition and with a volume over 300m EUR, is funding many startups at an early stage (Räth 2013).

The introduction of the UG ("Unternehmergesellschaft), similar to the British Limited Company (Ltd.), was also an attempt, to help a startup found its first entity without the necessity of dedicating 25k EUR for a GmbH – with a straight down payment of 12.5k – or opening a Ltd. in the UK. In addition, the German government, on Federal and State level, supports incubators at universities, regional advice & support programs for startups, events and initiatives. As mentioned, all these programs have a bridging function, however they can also harm the system by too bureaucratic processes and misleading incentives.

The under-financing of startups can be improved through governmental exemptions of tax on capital gains, in case investment companies, which invest in startups and small companies, are selling their equity. This aspect exist in both countries, Germany and the US, and is being extended in both countries (PWC 2013; The Whitehouse 2013). However, in Germany this tax break is limited to stakes below 10%, and the whole issue has been in an open debate for a long time. The last debate took place in 2013 between the government of a right coalition of the Christian Democrats and the Free Democrats in the German *Bundestag*, and the Social Democrats and the Green Party in the German *Bundesrat* with a majority of the left parties (PWC 2013). In the US, the Obama administration has been trying to get a permanent elimination of capital gains tax for amounts of up to $10m or ten times the taxpayer's basis in the stock (The Whitehouse 2013).

Even though the US is already considered a role model in terms of conditions and opportunities for startups, it continues to improve its government support programs and fosters the interaction between successful entrepreneurs, government and aspiring founders. On the website of the Whitehouse, one can read that "Startup America" is a White House initiative to praise, inspire, and accelerate high-growth startups and entrepreneurs throughout the US. It is an coordinated public and private initiative of the most innovative entrepreneurs, businesses, academic institutions, foundations and other leaders that work together to significantly increase the occurrence and success of America's entrepreneurs" (The Whitehouse 2013). The Whitehouse, under the Obama government, has also announced that it will commit $2bn over the next five years, as a match to private sector investment in promising high-growth companies through the Small Business Administration (SBA). $1bn will go to those funds that invest growth capital in companies, which are located in underserved communities. The other billion will be for an early-stage innovation fund, providing a 1:1 match to private early stage seed funds (The Whitehouse 2013).

The Startup America Initiative also tries to raise the tax deductions for Private Investment in Lower-Income Communities from 3.5 to 5bn USD. According to the Online website entrepreneur.com, it is almost impossible for individuals to receive grants from the government for founding a company like EXIST does in Germany (Entrepreneur.com 2009). In general, the site also mentions that the best chances for small companies to receive grants are for concepts, from which society will directly benefit or which are for non-profit. The US government website grants.gov provides a database with all current and post government grants in the US. It confirms the statement by entrepreneur.com. The other government website sba.gov underlines this fact with the statement *"The federal government does NOT provide grants for starting and expanding a business."*(U.S. Small Business Administration 2014a).

In the following, German programs for founders, investors and institutions will be displayed in further depth.

6.7.8.1   Government Programs for Founders in Germany

Various programs for founders in Germany can be found in the following list:

Scholarships:

Federal scholarship (in cooperation with a university)
- EXIST Gründerstipendium 'Founder's scholarship' (Exist 2013)

- One year support for up to 3 founders
- Per person: 800€ p.m. for students, 2.000€ p.m. for graduates and 2.500€ p.m. for founders with a PhD
- 17.500€ for material expenses
- 5.000€ for coaching
- Free office by university

- EXIST Forschungstransfer 'Research Transfer' (BMWi 2014b)	- Financial support in two phases - Phase 1 for 18 months providing salary for 3 founders and extra 60k € for material expenses - Phase 2 up to 150k € covering 75% of the costs of a the new company
State scholarships (examples) - Junge Innovatoren 'Young Innovators' (Junge Innovatoren 2013)	- Up to 2 years 50% payment of a basic salary at a public university/ research institution (TV-L 13) - 20,000€ for material expenses - 5,500€ for coaching
- Seed Stipendium 'seed scholarship' (Sächsische Aufbaubank 2013)	- Equivalent to EXIST *Gründerstipendium* - Without coaching and material expenses - Only for people residing in Sachsen
- Karl-Steinbuch-Stipendium (Karl-Steinbuch-Stipendium 2013)-	- Scholarship for students with a first degree (Bachelor or Magister) - Up to one year 830€ p.m.
Coaching & financial sponsorships **Federal** - KfW Gründercoaching - 'Founder's Coaching' (KfW 2013b)	- 4,500€ support for qualified consulting (within first 5 years)
State - Innovationsgutscheine - 'Innovation voucher' (Ifex 2014; Innovationsallianz NRW 2014)	- Between 2,500 and 10,000€ support for external research and development (50-80% of costs will be covered) - Various forms depending on the state
Financial credits, loans & equity **Federal** - Various KfW credits - Bürgschaftsbank 'Guarantee bank' - Wagniskapitalzuschlag 'Risk capital award'	- Low credit rates between 1.6-3% - Handling through private banks - For young and small companies up to 10 years existence and less than 500 employees - Up to 80% of the risks will be covered by the German *Bürgschaftsbank* if founders are

(Bafa 2013) **State (examples)** - L-Bank (similar to federal conditions)	not eligible for a credit at the terms of a private bank - The *Wagniskapitalzuschlag* offers the founders a 20% booster on an investor's investment. Up to 200k € of a 1m € total investment per year. Per investor 50k € max on a 250k € investment
Federal unemployment payment for founders	- Eligible persons for the German "Arbeitslosengeld I" can instead receive the same amount for founding their own company - Need to show a valuable business concept and convince the responsible officer at the unemployment agency - First 6 months with extra grants of 300€ p.m. - Next 9 months only 300€ grants
Support and coaching through government funded university programs	- EXIST III + IV program provides funds for certain universities to offer services like coaching, office space or external support founders from university

Table 36: Government and State Programs for Founders

From the experience of the author, the programs are in general a good start for helping young entrepreneurs, however they carry two major problems along. First of all, many of the programs like EXIST require a sort of employment position within a research or university institution, in order to receive the grants. This has several consequences; instead of being an entrepreneur with a limited budget for himself and his company, the founder is bound to the institution and its rules. According to many founders, it actually feels like being more an employee or PhD candidate rather than being an entrepreneur.

The intended protection of public money, by tying the startup close to an institution, rather slows the team down and does not prepare them well for being out in the market with their own company. This leads to the second drawback of the programs and their restriction is that it places the startup in the wrong environment, which has not many similarities to a business market. The result is that many teams with such grants in the end have to apply for continuing scholarships or grants (like the German program *Junge Innovatoren*), because they have not been serious enough about finding external financing outside the academic and public grant system. Among Ifex members (an initiative and institution of the State of Baden Württemberg to foster Entrepreneurship) this issue had been discussed several times in 2010, because the state of Baden Württemberg had received many of these applications for a continuing support through *Junge Innovatoren* investor (Ifex 2010). And even teams that had received *Junge Innovatoren* were in most cases not able to finance themselves after the end of the program or did not have found a non-public financing afterwards.

6.7.8.2 Government Programs for Investors in Germany

The four major government support programs for investors in Germany are the *Wagniskapitalzuschlag* ('investment add-on') (Bafa 2013), the investment matching through the *ERP Startfonds* (KfW 2013a), the VC fund increase program called *EIF/ERP Dachfonds* (BMWi 2014b) and the law for tax free revenues on startup financing and reinvestments (Bundesministerium für Justiz 2014).

The *Wagniskapitalzuschlag* gives the investor 20% of his or her first investment into a startup – up to a volume of 250k € and, therefore, providing 50K € – back without losing equity or the necessity to pay these 20% back to the government. This extra money does not give the investor more shares, but lowers the risk by eventually investing only 80% of the own money for the same valuation. Therefore, this government program is attractive for both, founders and investors, because providing this sort of bonus makes the investment financially more interesting, and thus increases the chances for startups to receive funding.

The *ERP Startfonds* offers a doubling of an investor's investment at the same conditions for the KfW as for the investor. It requires a lead investor setting up the terms, nevertheless the program has certain requirements and evaluates the startup team again through its own members. In the end, this program is also interesting for both parties, the investor as well as the startup, because it gives the investor the chance to distribute the risk also upon a another investor. The startup profits, because it can double the amount of investment, by concentrating on only one investor with the KfW following. In practice, however, the negotiations with KfW can also take a long time, diluting this effect of an easy doubling up of the investment sum (Author's experience with his own funding cases).

The *EIF/ERP Dachfonds* helps venture capitalists to increase their venture funds volume. It uses German as well as European money to accomplish this. The initiative is, however, run on a European level based in Luxembourg, and therefore cannot be seen as a mere German program.

The tax free and low taxes on investments in startups have been partly described in chapter 5.9.3.1. Besides paying no taxes in the rollover effect, there is also the case of paying just 5% taxes in case of a shareholders being a capital company that has more than 10% shares or significant influence on the companies it holds shares in (German law § 8b KStG).

6.7.8.3 Government Programs for Research Institutions & Universities in Germany

The *BMWi* ('the German Federal Ministry of Economic Affairs') has several startup initiatives running. Most of them are part of the *Mittelstand* chapter, referring to mid-size companies underneath the innovation section, others can be found underneath the education chapter.

The EXIST programs (EXIST I-IV) are an effort to improve the entrepreneurial environment at universities. Altogether, in early 2014 the website of the Ministry of Economics Affairs listed 41 initiatives for fostering entrepreneurship at different educational institutions located in different states (BMWi 2014c). The *BMWi* evaluates its programs on a regular basis through the use of external consultancies or institutions like the Fraunhofer Institute. Besides the obvious positive effects of increasing the awareness and the positive will, there are on going discussions between investors and entrepreneurs about these programs and if in the end they really help or are even being hindering by increasing bureaucracy or are distorting the startup's market reality (Blumberg 2015).

6.7.9 Mixed Sub-Element: The Founding Process

The founding process is a mixed sub-element partly belonging to the Culture, the Business, as well as the Government Element. It is an important step, as described in chapter 3.6.1, because it separates the thinker from the doer. Many people like the idea of having their own company; they play with the thought of being an entrepreneur, but only approach the actual founding "asymptotically". In parts, Germans might be more hesitating, not only because of the uncertainty and risks, but also because of the workload and bureaucracy connected with founding and running a company. Institutions like the "Arbeitgeberservice" ('employer service'), the employment center, incubators and programs in Germany actually try to support and accompany potential founders through this phase. However, there is still a comprehensive service missing, mainly the right information and a checklist, what steps are necessary in what order. Approaching the topic from a different perspective, one also generally intends to help the founder starting a company, by offering scholarships and initial funding. The German EXIST *Gründerstipendium* program in Germany is a typical example for it (explained in chapter 6.7.8.1). Ultimately, it is important to clear the way of unnecessary hurdles for potential founders, in order that they can gain some momentum. However, the founders should not be misled, by pushing them into a founding or releasing them from too many hurdles. Founding and running a company is accompanied by a constant flow of problems and many things not running at the speed or accuracy as planned. Being in a too pleasant easy environment and padded by scholarships etc. might ill prepare the founders, for what is waiting ahead of them.

6.7.10 Social Networks

Social networks are an important source for entrepreneurs to connect with the different elements of the Entrepreneurial Ecosystem. Whereas in the inner circles social networks like Facebook play more a point-to-point role between the founder and a few individuals, in Circle IV, the founder can reach out and promote his startup to a whole community and customer base. It is not only the theoretical proximity and speed to millions of customers, but also the opportunity to use technology, platforms and help of the *crowd* to enter new markets.

Today, there are physical social networks like alumni, business clubs, as well as virtual ones like Facebook or LinkedIn. The importance of the alumni clubs has already been described in the previous section, however these new virtual social networks have an immense influence and impact on the entrepreneur's life. On the one hand, they are the online extension of these physical networks, while on the other hand they at the same time break some boundaries, open networks and contacts. Many people have today online contacts, which they have hardly seen in the "real" life. Nevertheless, yet they can be of service and help their business. The networks like Facebook are also a platform for business itself, bringing customers, contacts and friends together. An essential part of today's online marketing occurs and is influenced by these virtual social networks. In addition, these networks can help in finding new employees, get instant feedback or make first sales. Eventually, they are a free communication platform and facilitator for many other aspects. With the international community in each of these big networks, entrepreneurs can reach almost any person anywhere and at any time. Virtual networks reduce time and complexity for managing and remaining social contacts; nevertheless, the pressure may increase due to the expansion of contacts and connections. Studies used to say that human beings can manage up to 150 contacts thoroughly (Dunbar 1993)(Daily Mail UK 2010). Many well-connected entrepreneurs and persons today might have, however, between 500 and 2000 contacts online. This leads to a new challenge for these individuals to make proper use of this high amount of connections and to stay focused, besides the many distractions coming from these high volume contacts.

6.7.11 Co-Founders

Each co-founder has his or her own ecosystem and merges it with the other co-founders. Since resources and money are rare elements for founders, they need to strengthen the core of their team with co-founding partners, thereby trying to compensate the fact that they cannot choose the best employees from the market to complete their team. That is why co-founders are the most important part of the team aspect. They are the base of each company, whether it was the co-founders Bill Gates and Paul Allen from Microsoft, Larry Page and Sergey Brin from Google or the Samwer brothers from Rocket Internet. Often one of the co-founders stands more out than the others in terms of PR, and sometimes the co-founders are not even known, even though they played an important role in the background. Especially in the early founding phase, as well as in difficult times, co-founders can motivate each other and bring in different perspectives and solutions. In terms of the Entrepreneurial Ecosystem, the Co-founder Element ranges between the most inner circles (Circle I+II) being a friend, who influences and inspires the other founder's skills, through the Educational Circle (III), in which the founders often get to know each other and team up, and to the outside of Circle IV, where they apply their expertise and launch their business together.

6.8   Business Model

The term business model is much discussed in theory and practice. For the layman the business model is often equivalent with the revenue model – and many startups have initially the same understanding. However, experts from academia or the VC industry see much more to it. Al-Debei and Avison (Al-Debei et al. 2008, p.7ff) see the business model as an abstract form of an actual company with all necessary elements that are needed to fulfill the company's goals and objectives. This includes, besides the revenue model, many aspects like the value proposition, the network or the key products. For George and Bock (2011) a business model is the design of an organizational structure, which is needed to enable a business opportunity. For Osterwalder et al (2010) it is about creating and delivering value, not only in an economic but also in a social context. The group around Osterwalder introduced in 2010 a graphical version of the business model, they called the "Business Model Canvas". This canvas pictures the various fields that are needed for a company to create and maintain value. Many startups but also large corporations are said to have used the canvas model to design, question and improve their business model.

The main elements of the business model canvas can be divided in four areas:

1. Value proposition
   - Customer value, demand
   - Service & product offering

2. Customers
   - Customer relationships
   - Customers segments
   - Channels

3. Financials
   - Cost structure
   - Revenue stream

4. Key infrastructure & processes
   - Key resources
   - Key activities
   - Key partners

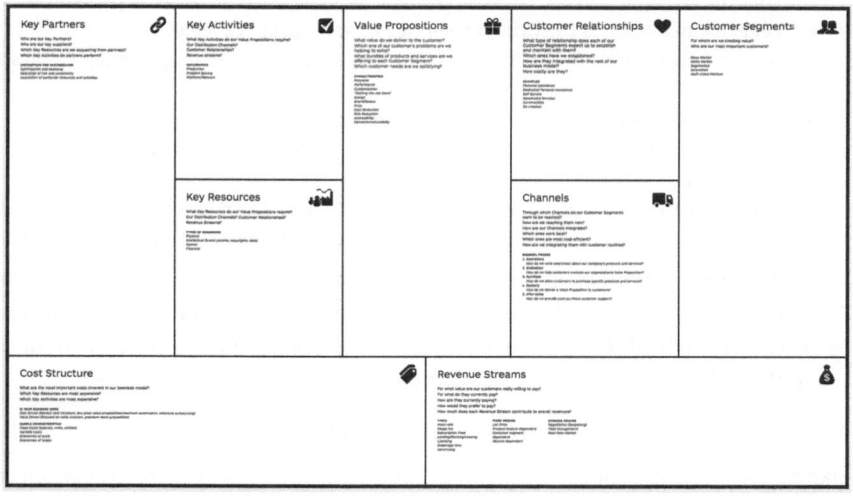

Figure 25: Business Model Canvas (Osterwalder et al, 2010)

The business model depends on the Entrepreneurial Ecosystem of each country or region, and therefore must be adapted by the entrepreneur. A key contact to a champagne vineyard, giving the entrepreneur exclusive international distribution rights for all Arab countries, might not be of high value, because of the prohibition of alcohol in those countries. Another example would be Facebook or Google, which are mainly blocked in countries like China, and thus their business model does not work there. All circles and elements may have an influence on the business model, like the skills of the entrepreneur helping the startup to build up a unique product, or the network from college that gives access to particular key partners. How to structure and develop a business model, is something that can be taught in the Higher Education Element. Nevertheless, the Public and Business Circle and its elements have the most concrete influencing factors, because of the direct connections with building up the business itself. Important aspects are:

- Market maturity: Is the market already mature enough for the business model?
  (e.g. selling mobile applications on a market with a low smartphone density)
- Business conduct: Are business partners able to work with the company?
  (e.g. cloud services that cannot be used at large corporations due to compliance rules)
- Cultural & religious acceptance: Do many customers accept the service?
  (e.g. beef products in India; escargots outside France; capsule hotels outside Japan)
- Legal frameworks: Is the product, service or sales practice legal?
  (e.g. products with a medical promise but no license, lending money in Arab countries)
- Regional fit: Do persons in other regions also like or need the product?
  (e.g. beach clothing in cold areas, building material like wood for houses in Germany)
- Language: In which language do customers prefer to communicate?
  (e.g. expanding an online marketplace in English to France that prefers French)

With the many involved aspects/ elements of a business model from the Entrepreneurial Ecosystem, the interaction of these elements becomes crucial for the entrepreneur. Only if he or she understands the important elements and their connections, the right business model can evolve.

6.9   Interaction within the Entrepreneurial Ecosystem

Like any other system, the Entrepreneurial Ecosystem also lives from the interaction of its elements (compare Chapter 6.3.), and the success of a founder and the startup always relies on several factors, which influence, hinder or support each other. Like in Horvarth's (1937) "Der Jüngste Tag" the final results are always a combination of many different incidents, which are not predestined and also open for potential for change. Many young teams, from a technical background, for example believe that the uniqueness and specialty of their product is enough to be successful, unfortunately, they often never get close to selling a single unit of their product. For a startup, this could also mean that despite a great team, a strong investor and promising product, it still may not be successful, because, for example, the market was not ready. On the other hand, a team without the financial support of an investor might have a too slow market penetration, thus competitors could pass by them.

Using the language example again, a person will never be able to master a language properly, no matter how good the abilities and skills are, if there is no constant almost daily interaction with the environment. If there is no occasion to use and speak the language, there will always be something missing. The same is the case for entrepreneurship. The perfect theoretical concept needs to become applied in the market, in an interaction with the elements of the ecosystem. If the market, for example, is missing, the entrepreneur's concept will remain a theory. It must be remembered that all the elements are influencing each other in different ways. In order to master the language or the business, a certain mixture is needed. Professor Schwarzkopf's *Hand of Fortune* is figuratively describing this idea.

6.9.1 Hand of Fortune

The hand of fortune, used by Schwarzkopf (2010) in his business law classes, is a very good illustration of these necessary combination of success factors. Schwarzkopf mentions five elements to be crucial for any success and pictures them according to a human hand:

- Thumb standing for **Character**: The first and thickest finger stands for the character of a person. Like at the hand, character comes first and defines or sets the base for each activity of a person or entrepreneur. This matches the setup of the thesis' Entrepreneurial Ecosystem that defines the character being part of the first and most inner circle.
- Forefinger standing for **Experience**: The forefinger stands for experience and the ability to deal with and learn from it. Since the forefinger is used in many cultures to point on things or raise and shake it as a form of warning and teaching, it also reminds the person not only to learn from its own but also from the experience of others. Without this ability to learn and use the experience, the entrepreneur might make too many unnecessary mistakes or does not improve enough. Experience is also part of the inner circle (Circle I) of the Entrepreneurial Ecosystem.
- Middle finger standing for **Knowledge**: The longest finger stands for the knowledge and expertise of a person. The length and middle position emphasizes the central role for success and the depth required. It therefore describes the content part of what is being done. With the description of the first three fingers, Schwarzkopf is quasi completing the Personal Circle of the Entrepreneurial Ecosystem. With three out of five fingers from the Personal Circle, Schwarzkopf clearly puts the emphasis on the entrepreneur, and not declaring the environment to be mainly in charge of success. However, the fourth and fifth finger add also other aspects.
- Fourth finger standing for **Connections**: The fourth finger is the sign for connections and partnerships. Without the bindings and cooperation, success will be hard to reach. Every step alone might take too much energy and make the person too slow. Furthermore, it also points out the relation with customers, suppliers and employees, which are essential for any business. This description

matches part of the Personal Circle (Circle I) as well as part of the Public & Business Circle (Circle IV).

- Small finger standing for **Good Fortune (Luck)**: Eventually, the smallest finger brings in the element of luck and fate. It is, however, not only a passive form standing for the things we cannot influence. It also invites a person to deal with the circumstances and to be aware of opportunities, which require action from him or her. Staying within the framework of the Entrepreneurial Ecosystem, good fortune supposed to be part of the rules and development within the system and thus can affect all elements in all circle areas. It could range from being at the right time and place to meet an investor or co-founder, to catching the right momentum of a market that is ready for the new business concept.

These elements, like the fingers of a hand, function the best if they come or are together. Any finger missing will limit a person to use the hand properly or fulfill all tasks with it, up to the point, where it is impossible to succeed. The same is the case for success in business. The mentioned elements altogether provide better chances for success. It is possible to be successful with an element missing, but it is less likely or more difficult. The lack of the small finger – luck, fate & opportunity – could ruin the best team, despite its great contacts, experience etc., when the market opportunity does not show up, or when simple an external event is changing the situation (e.g. market break down, stolen IP, destroyed product, new technology, insolvent supplier, dropping out of key personnel). On the other hand, the luckiest team lacking the "knowledge & expertise finger", will sooner or later fail, because it lacks the core substance, for which the company stands and how runs the business. This will lead to poor quality, low innovation or a slow market penetration for example. Failing could also come from competitors taking over market shares quickly, when the team is too weak and has no expertise, knowledge or information (middle finger), as well as experience (forefinger) to hold the ground against the competitors. This includes of course the necessary management skills to run the business properly.

The broader the fingers are defined, the more they can be used to summon almost any use case or business venture and reflect their success chances. The sharper the fingers are defined, the more specific the cases can be analyzed. Schwarzkopf's hand of fortune could, therefore, perfectly be used in two phases to evaluate a team and its business: First, in general for a quick overview and second, when scrutinizing the elements in depth. An issue tree (or logic tree) would be a matching form to cover those two phases. It is a common tool used especially in management consulting, in order to cluster and solve problems. In this case, the tree will always have five branches according to the hand of fortune, which defines the first stage. The second stage or phase will then build sub-branches in dependence to the team and finger element being analyzed.

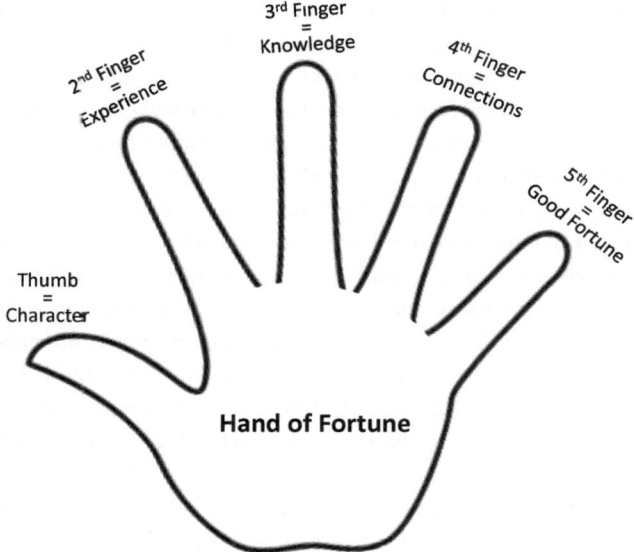

Figure 26: Hand of Fortune

Thumb	**2nd Finger**	**3rd Finger**	**4th Finger**	**5th Finger**
Character	Experience	Knowledge	Connections	Good Fortune
Personal Circle Elements			Private, Public & Business Circle Elements	Laws and connections within the ecosystem
Examples				
- Motivation - Endurance - Visionary - Ethics	- Industry experience - Startup experience - Management experience	- Coding skills - Project management - Technical knowhow	- Customer access - Investors - Partners & Supporters - Family & Friends - Institutions	- Meeting a client by coincidence - Entering a market at the right time - Receiving money before a crisis - Crisis creating an opportunity

Table 37: Hand of Fortune - Characteristics

6.9.2   Interacting through Open Innovation

Another form of interaction can be described through open innovation (compare 3.8). Besides the regular form of cooperation that requires several elements from the ecosystem, the entrepreneur applying or being in the middle of an innovation process has to understand, which important elements are involved and under what circumstances. Ecosystems will overlap, when several entrepreneurs are engaged, and the rules how elements of each ecosystem interact might be adapted. Investors might be more hesitant, when IP is split among several entities, and a USP towards customers could be difficult to sell, because of the competing entrepreneurs with the same invention or product. In addition, the entrepreneur's skill set is also challenged, because a new form of social interaction occurs, with a group of participants that are neither customers nor employees or investors. With the use of the Internet, open Innovation can be world wide, expanding the influence of culture, language or geography dramatically. Open innovation, therefore, plays a strong role in the Circle IV. Furthermore, co-founders and social media are coming to the forefront. Through open innovation, new co-founders can be found, and social media allows ideas to be spread quickly, including the technologies that come along. Sharing systems like Googledocs or Dropbox as well as online collaboration tools like Podio make open innovation also feasible for startups and small companies that cannot afford expensive and sophisticated IT systems. Open innovation is, therefore, not an exclusive option for large corporations anymore, but is available for a variety of companies.

6.10 Important Phases and the Role of the Circles

The different phases a startup usually experiences lead to different ability- and skillsets, required from its founders and management team. This reflects very well the importance of the most inner area of the Entrepreneurial Ecosystem – the Personal Circle. In the pre-seed and seed phase, a lot of inspiration and braveness are necessary, in order to really set the founding process in motion. The closer one moves to the market entry, the higher the requirements become, such as, for example, communication skills used to talk to customers, employees or partners. The more complex the product or service is, the more the founder needs be focused, for preparing the product for the market and thus first sales. Once external investors become involved, the founder or the team must bring some financial skills to the table. In the "early stage" and "growth phase", endurance and self-motivation are very important, because for most startups immediate financial resources or enough money are not available, to steer the team through these rough times. The more the startup matures, the more management skills are required. Once the first customer base is settled, the market penetration requires a structured and stringent approach. Constant improvement, adaption and management of limited resources become crucial. Once the company is set in motion, the entrepreneur's ecosystem should be functional and operating in all circles. The Private Circle needs to backup the founder, especially in the first years, when the financial existence is in question. Essential advice arises out of this Private Circle, where friends and family give often their direct and blunt opinion – without even being asked. The Educational Circle should have taught the skills and expertise for the startup to get ready, and it often provides the first co-founders, supporters and employees. The more difficult the product development and closing of first customers becomes, the more professional advice from Circle IV, the Public & Business Circle, is required. Once on the market, Circle IV plays the major role, because it is part of the business and of the market. The ecosystem in the founding stage becomes for the first time fully functional on all levels – even though it is far from complete.

6.11 Deriving Key Elements for Improvement

Within the Personal Circle, it is mainly the skill element that can be improved by extensive training and through experienced entrepreneurs. This fact was also stressed in this thesis' survey by investors and by the entrepreneurs as the top two success factors. When observing the next circle, one recognizes all of its elements consisting of real persons. Thus, improving their influence on the entrepreneur is hard to identify and implement. A changing culture and more successful entrepreneurs and business angels could lead to more coaches in the Private Circle that have an influence on entrepreneurs. In this circle, one finds less improvement than in Circle III, because the Educational Circle offers greater potential, because it deals with education and training, which of course can be analyzed, changed and improved. In order to change or improve elements like the culture of Circle IV, entrepreneurial education already at high school has a potential for a positive influence. Circle IV certainly has many other elements that bear potential for further and different improvement. Aside from culture, a media push to promote entrepreneurship is required. Furthermore, solving the language issue for German and European startups, which common language to operate with, would help to expedite the growth faster in other European countries. Government improvements for better tax and bankruptcy laws are another two examples for improving the Entrepreneurial Ecosystem. The comparison between the US and Germany has also taught to think big in terms of geography. Pushing the European idea further will certainly create a bigger and more homogeneous market for entrepreneurs and established companies. Business- and scenario planning must be constantly modernized and partly adapted to the US approach. These aspects will be considered in detail in the following chapter.

7 Improving the Entrepreneurial Ecosystem

When improvements are considered, one has to distinguish between the direct measures and influences and the indirect ones, which are hard to change by governmental regulations or new measures. Circle I and II of the Entrepreneurial Ecosystem are belonging to the latter group. Within these circles, it is role modeling and the general positive environment that have a determinative influence on the entrepreneur. The only exception to this is the Skill Element, which is directly influenced through better training and up to date courses, although the measurement might come from a direct change in the Educational Circle and its elements, such as at the university level. An indirect improvement can come from the Experience Element, which can be improved, if one had more serial entrepreneurs and startups to work directly with the founder. He or she can potentially relate and transform parts of the serial entrepreneur's experience to him or herself. Circle III and IV can often offer numerous suggestions for direct improvement. Nevertheless, entrepreneurship will always be driven by the entrepreneur and can only be supported and accelerated by the other elements of the Entrepreneurial Ecosystem – from Circle II to IV. Just like a car needs an engine to move, this ecosystem needs an entrepreneur to function. Throughout many research initiatives, entrepreneurs are becoming more and more a business object, which is being scrutinized for an understanding and optimization. Nevertheless, one must not forget the personal aspect; otherwise the risk of misplacement or false categorization might occur. Many entrepreneurs have told the author that they feel sometimes like a guinea pig, whose work, attitude and motivation are surveyed and questioned almost every week. The whole optimization process should not be exaggerated to the point, where it literally kills the object – the entrepreneur. Now having said this, the following concepts and approaches try to enhance, foster and change the ecosystem for an entrepreneur, in order to have more latitude, freedom and support for building up successful ventures.

7.1 Fostering Entrepreneurship on a Higher Educational Level

The circle, where direct measures could be first undertaken in, is the Educational Circle. Here, the entrepreneur receives his skillset, becomes hopefully further inspired and is systematically brought in touch with elements of Circle IV.

7.1.1 Integrating Experienced Entrepreneurs

Many universities have started to either invite entrepreneurs and top managers for holding lectures or have hired them, for example, as visiting professors. This type of activity and integration needs to be further developed, and the sparring partners and coaches for the aspiring entrepreneurs need to be wisely picked. Whenever and wherever young founders are theoretically taught the elements of entrepreneurship, practicing and successful entrepreneurs should be present. This approach is not limited to education, but also relevant for business plan competitions, scholarship programs or consulting centers for entrepreneurs.

Experienced entrepreneurs tend to be more open and cheer and motivate the aspiring founders, instead of being too critical – especially at the first encounter. They can always bring in examples from their own company and make concrete assessments and recommendations, as if they were starting the new company. The non-entrepreneurs warn the aspiring entrepreneurs often about founding a company, risking everything and living for years on the level of a welfare recipient. Many people, especially in Germany, believe it is always helpful to stress the problems and risks, in order to avoid mistakes. This form of pessimism and fear of failure already at an early stage can, however, hinder entrepreneurs to get started in the first place and must be changed. A mixture of entrepreneurial advice on the one hand, and analytical partly critical support from other involved persons on the other hand, could be the solution for the German system.

Involving these experienced entrepreneurs could also help to turn them into coaches of the founders, and thus becoming a part of the inner Private Circle, thereby having an even stronger personal influence upon the founder.

7.1.2   Flexibility in the Student's Curriculum and Part-Time Employment

Most startups need a certain momentum, time for testing their idea and critical speed to proceed, thus the curriculum needs to be flexible for the students and their startup development. Pushing students to quickly finish their degree, in order to start working earlier and even shortening their time abroad – studying or doing an internship – is wrong. There needs to be a counter-motion, to slow down the academic career path, by giving more flexibility and time for trying out and playing with different business ideas. In addition, time abroad is also a part of the maturing process of a student, which helps not only the entrepreneur, but also the employee to have more life experience before starting to work. These observations are also expressed by many recruiters and employers (Preuß, & Osel, 2012; Himmelrath, 2012).

If a student works part-time on an idea or project, which could be turned into a startup for example, the curriculum could help to further his or her aspiration. On the one hand, academic papers, seminars or master and doctoral theses could be used, to work out in detail some parts of the potential venture. On the other hand, the student needs to be able to take off a few months – a sort of startup sabbatical – to reflect and to test the concept, in order to get feedback from the market or to run a prototype, for later use. If he or she then decides not pursue or postpone the founding of a company, it should be possible to continue the academic. Postponing exams or papers to a later semester need to become possible. Another way of support are part-time jobs at university after graduation, which provide some financial means during the time of the development and evaluation of a business idea. In addition, a part-time job helps to give the graduate student a social net, which is of great support – especially in Germany – because many people fear the fact of having no job after their graduation. The Karlsruhe company PTV is a good example of a company, which had been founded and operating during the part-time jobs of its founders at the university – where they had been working as research assistants (Stucky 2013).

7.1.3 A New Form of Business Planning

Depending on the scenario and on the projection, the startup needs to mirror the necessary actions and resources in the business plan and vice versa. Aside from

- difficulties of the business planning itself;
- wrong and demotivating aspects coming along with application of a business plan;
- complex structure, as well as the timing, when the plan has to be written, and
- case modeling, which continuously needs to be adapted and improved,

an improved business planning concept and process could be of great help for the founders. The word business plan should, therefore, be understood more as the planning of the business rather than the design and completion of the plan itself.

An improvement of the ecosystem could be, to develop a new business plan model in terms of timing and in accordance to the requirements of the market, and not the requirements of universities or competitions.

7.1.3.1 Golf-Layout Business Plan

Rather than asking the entrepreneur to write a business plan, which covers the essential business elements in a more or less chronological order, one should develop the next step for the entrepreneur individually, leading to a sort of step-by-step "field approach", which gradually considers the next crucial element. A founder with a high implementing capacity, but low self-esteem in terms of founding his or her own company, should be motivated by finding a great market demand and by understanding that founding a company is not as dangerous as jumping with a parachute out of an airplane. If this type of founder was to write a classic business plan right from the beginning, the challenging steps to be covered in that plan would often lead to more doubts about the business and probably to give the venture idea up.

A technical expert on the other hand, who has been developing his prototype for a long time, and who is convinced that it is already the greatest product, should rather be focusing on understanding the market and what the customers really want. If that person was to write immediately a business plan, the planning of sales, for example, would often be over- estimated, aside from the fact that writing the plan would take too much time.

Other typical examples are strong business oriented founders, who often understand the market demand better, but lack technical implementing capacity. For them, it is critical to think of the solution they want to offer, the resources they need and how to fit them into the team. Furthermore, they need to consider, how to afford them – either by finding cofounders or by hiring employees.

The on the spot "field business planning" can help even the inexperienced founding supporters, to add value to the entrepreneurs. The field approach in parts can be compared with the layout of a 9-hole golf course. The 9 holes represent the important elements of a business plan. This layout would allow the player to play the holes in a non-chronological order, thereby contradicting the standard golf rules:

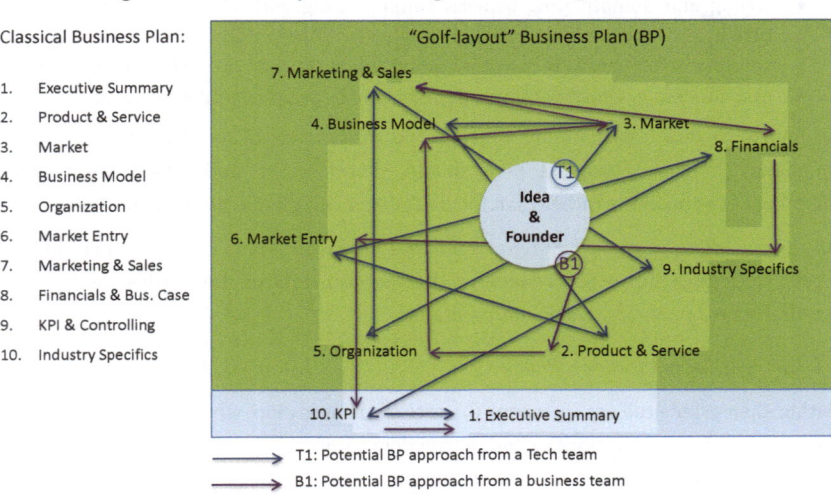

Figure 27: Field or Golf-Layout Business Plan

The classic business plan approach would let the player play one hole after another – despite the fact that, according to the executive summary, hole 1 should be played last.

The field approach instead will allow the entrepreneur to choose the hole or topic, which at his or her stage is the most important at that time.

7.1.3.2 Application by Different Teams

Following the example from above, the business-oriented team might look into the product development and the necessary resources and personnel for the organization. The tech-oriented team on the other hand should rather scrutinize the clients of established competitors in the market section, in order to get an impression, of what and where customers are currently buying. If the team is still convinced that their product could be a huge success, they should further try to get direct feedback from potential customers on their product idea, and then develop a business model around it and identify the necessary resources. With the business model in place, the team could then observe the market and question the initial thought, which they, as a business team, usually have started with. If the business model matches the targeted market, they could continue with the necessary marketing and sales structure, before calculating the business case in the financial section. A tech team, on the other hand, after having developed their business model, might match their initial product ideas with the business model and after that discuss the market entry. If they are convinced having identified suitable market entry point, then they could start calculating their first business case. The financial contemplations might lead them to thoughts about their organization, especially about the necessary team for their business and the location, where they want to develop and/ or produce their product. The financial thoughts should call their attention to the necessary marketing, sales structures and upcoming "ToDos".

Eventually, they need to think of industry specific aspects, such as patents, certificates, administrative regulations or state laws, which could hinder or slow down the implementation of the business idea. Even industry standards like payment conditions could be of great influence, when late payments are common and advanced payments are not, leading to potential cash flow problems for a not thoughtfully financed startup.

The KPI elements could be considered the driving range, where the entrepreneur can practice and challenge many of the main ideas, questions and derive the key performing indicators (KPIs). Examples for these KPIs and key questions could be:

- How many customers are needed with the current pricing to reach a breakeven?
- What conversion rate is common in the market, and is that matching the estimations?
- What are the revenue streams? Are they sustainable, and how can they be adapted?
- How much should the production costs decrease per year?

After having calculated the business case, the business team could check, whether their business idea requires certain industry standards – for a website idea this might be certain laws for data protection or for a software team potential SAP standards. This could influence their market entry, to the extent that the team might re-evaluate their entry options. Having covered many of the possible fields, the business team can before ending in the executive summary, summarize the most important factors for the founding and early stage phase in the KPI section of the business plan.

In order to collect money, entrepreneurs have to successfully demonstrate that they are capable of "playing the different holes", but need not necessarily have played all of them. That is why it is sometimes said that the best business plans fit on a napkin, showing the potential of the idea, the capacity of the team and the momentum of the business (Fisher n.d.). Which basically confirms the author's observation that the key starting aspects are the founder's motivation in alignment with a certain business idea.

Aside from this relative new understanding of planning a business, young entrepreneurs should be advised, how to develop a business case in Excel. A skill set that could perfectly be taught at university.

Another helpful add-on and part of the business case could be a new form of scenario planning, which will be explained in the following.

7.1.4   Scenario Planning in Business Cases

When business cases are being calculated with integrated scenarios, entrepreneurs often make mistakes. On the one hand, case calculation cannot be transferred from the origin and classical usage in computer algorithms (Cormen 2001, p.25ff). Many times, best and worst-case scenarios are being used in such way that a fixed percentage of a medium case is added or deducted (Anderson et al. 2012, p.142ff). A new company, however, has no history of revenues, which can be used for a base and thus medium calculation. Adding or deducting percentages for the best and worst case from this base value, for a better understanding of the startup's potential development, is therefore not possible.

7.1.4.1   Worst, Medium and Best Case Scenario

Once the first version of a business case is set up, scenarios can be used to simulate different outcomes, which is especially interesting for a startup without a post record. In practice, worst-, medium- and best case planning is then being applied.

When looking at this scenario planning from an investor's, as well as an entrepreneur's perspective, worst, medium and best cases should be calculated differently. The author, therefore, suggests a new form of scenario planning for startups:

- **Worst or surviving case**: The founder can only "survive" in this scenario. The money for further investment, growth or even dividends for the investors cannot be generated. The salary level for the founders is at a base level. Anything underneath that point will lead to the failure of the project. This is important for the founder, because it gives the startup the sense, what they need to achieve for themselves, in order to bootstrap (finance the company themselves) the company. It also indicates, what level needs to be trespassed, in order to attract and later satisfy the investors. The surviving case is also very helpful for startups, which are insecure, and regardless whether they need external investment or not. Many Internet cases might, for example, realize that the necessary sales needed for surviving are so high that they cannot continue without venture capital. This is the case for most business models, which need a network effect, in order to reach the profit zone. The network effect requires a huge number of initial users that often have to be bought expensively through marketing efforts, which usually no startup can afford. If a startup wants to bootstrap in this case, they either have to finance their base operations and experiment reaching the network effect without heavy investments, or they hope something unexpected will happen. Hope is not the best advisor for a startup. From an investor's perspective, this is certainly not the worst case, but more like the failing case, where they cannot get their money back. Romans (2013, p.107), therefore, suggests the worst case from the investor's perspective to be, when the investment sum can only be paid back without dividends or interests.

- **Medium "comfortable" case**: This scenario makes the founders feel comfortable. After a certain time, they can pay themselves decent salaries; they can invest in growth and eventually pay their investors dividends including a standard return on invest with low interests. Another situation in a medium case could be that the company becomes attractive enough, to be sold at a standard market evaluation. This is the typical German way of calculating a conservative scenario. Due to the limited time and few skills operating with a tool like Excel, most startups do not have a second or third scenario. They often end up with this medium case scenario. As most investors have told the author, they hardly look into the Excel business cases, but rather interpret the KPIs and numbers presented when making a decision, whether to invest or not. This demonstrates, how important this medium case and the listing of the KPIs in this case have become. However, what a medium case often does not show is the complete potential of the startup and its business concept. Generally, investors bet on potential and the ability to seize it. Therefore, a scenario is required, which shows the full potential of a business. In most cases this would also require a global estimation.

- **Best "Skyrocket" case**: This is the scenario, in which the startup performs to its full potential, becoming a global player, and thus gaining a substantial share of the serviceable market. Not that this scenario is more realistic, it simply shows, in which financial area the business could range. Here the investors can better judge, whether it is rather a 10, 100 or 1000 million dollar business. Big VCs often only invest in startups that have the potential for a 100m+ exist or even higher, thus being able to generate the total amount of the fund volume (Romans 2013, pp.62–64). The figure 27. Above from the Airbnb pitch deck has shown, how the founding team can use

a) total available market,
b) serviceable available market and
c) seized market,

to show the complete potential and still calculate conservatively, when it comes to the seizing part. However, it gives the investors the chance to contemplate, whether they could help to seize more of the market potential. If the potential has not been shown to its full extend, the phantasy for entering a huge market is limited, and the market share, a startup needs to acquire in this smaller market, becomes relatively high. Investments are generally connected to potential and phantasy, because it is a bet on the future – with a team, a concept and a market that have to prove themselves.

7.1.4.2 Scenario Comparison

The following graph shows the three scenario types in a revenue-cost relation:

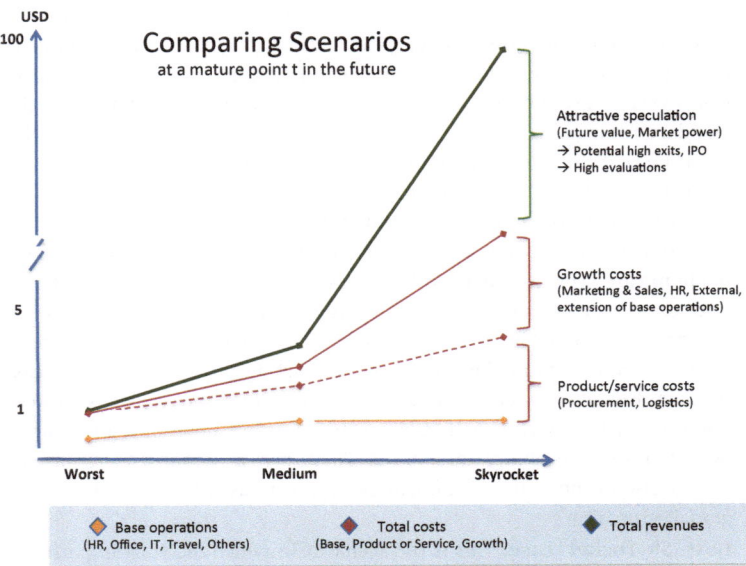

Figure 28: Scenario Comparison

In the worst-case scenario, the company can break even, if the revenues can pay for the base operations such as HR or office expenses including the product or service costs, which are related to the revenues they have generated.

The "medium scenario" has only a small increase of the base operational expenses, mainly due to an increase of the salaries for the team, which in contrast to the worst-case scenario can pay decent salaries. It will also bring an increase for the product or service costs, due to higher revenues, however its gradient is supposed to be smaller than the gradient of the revenue curve, due to economies of scale and learning effects. In addition, there will be extra expenses for growth such as marketing or extra personnel that exceed the base operations. Especially startups need this kind of growth money, in order to accelerate and overcome the hurdle of being unknown in the market. The medium scenario delivers some profit that can be used for dividends and interests for the stakeholders.

The "Skyrocket scenario" shows exponential growth in revenues, yet the growth costs also increase dramatically. However, the gradient of the revenues must be higher than the total cost gradient to stand for a profitable scenario after all. This gap and thus the profit justifies high evaluations at the several investment points, because it leaves plenty of space for speculation on the future value and market power, which could result in an IPO or profitable exit.

Maybe the "Skyrocket scenario" shows the biggest difference between US and German startups. It summarizes the attitude and situation of the German venture and startup players. The Skyrocket case should become the normal case level, having the leverage on seizing parts of a market. In addition, the ecosystem needs to be ready for such a global perspective. The elements must support the startup in thinking big and global, otherwise these international growth thoughts of German entrepreneurs might be doomed and drawn into a self-fulfilling prophecy of thinking too big and losing in the home court of Germany.

7.2   Comprehensive Founding Material and Information Base

The existence of many different supporting programs, requirements for a company foundation, as well as the different laws and obligations slow down the founding process and are often a huge extra hurdle for founders. These hurdles often lead to general doubts about the founding itself. Instead of concentrating, how to enter markets and gain customers, founders often are worried about tax reports, written warnings and other legal mistakes they could make. In addition, much time is spent on paper work, finding the right insurances or understanding, which official documents and registrations are necessary. Examples for that in Germany are getting a *Betriebsnummer* ('standard company number'); the search for the right *Berufsgenossenschaft ('employer's liability insurance')* and *Vermögenshaftpflicht* 'Professional liability insurance'; understanding the obligations when hiring interns or employees and writing the proper terms & conditions, as well as imprints before going online with the company website. A simple yet very effective way of helping to overcome these hurdles, would be a comprehensive information website with a founding checklist, all relevant sources and existing federal, state and regional support programs, as well as institutions. There are currently several websites that carry some information, but they are far away from being comprehensive. And as long as they are not comprehensive and approved by the government, founders will always need to do further research or hire professionals, in order not to miss an important rule or regulation. An improvement in this field requires the alignment of the Government Element with Media & Information Element of Circle IV.

7.3   Government Enforced Programs

Governments that want to improve their Entrepreneurial Ecosystem need to stimulate and honor the stakeholders of the system – from the startup to the investor. Tax deductions, the reduction of bureaucracy and opening the labor market are "freeing" efforts that aim on loosening restrictions. These above efforts are directly targeting founders and investors. On the other hand, forcing successful companies to invest parts of their profits into the venture scene can be seen as an "urging" effort involving elements and stakeholders of the Public & Business Circle that have not been a part initially of the Entrepreneurial Ecosystem (this point will be explained in the next section). Yet it shows, how elements are becoming actually a part of it, especially if they are coming from an economy and business ecosystem, which is growing or is a part of the Entrepreneurial Ecosystem.

7.3.1   *Corporate Venture Law - Increasing the VC Volume out of Annual Profits*

The government can help to increase the amount of risk capital in the German market, by introducing a new law for large corporations to invest at least 1% of their profits into venture capital business. This will lead to:

- A direct increase of venture money in general
- More money for existing and new ventures funds, which manage the VC investments of some corporations, if a company does not want to open its own fund
- More corporate VC funds for companies that want to manage their own VC money
- Employees & shareholders becoming aware of and probably interested in the startup and VC market, because their company invests into that market (directly

or indirectly). This will also support crowd funding platforms, due to the increased awareness and potential interest of employees in personally investing in startups

- Better access for startups to large corporations, because of the new awareness of the employees
- More media coverage, because of the increasing volumes and stories to write about

First approaches into that direction could be seen in France (Interviews with several VC partners), when several investment partners of VC have reported that under the Sarkozy administration negotiations between the government and big French companies had started about a partial investment of profits into risk capital. Some stories go as far as saying that Sarkozy wanted to force the companies avoiding an official law of 2% by investing 1% of their profits voluntarily. Without the government initiative, a variation of this was happening in the US and was a starting point for the growth of VC business in the US such as the investment in risk capital by pension funds. These pension funds (as described in chapter 4) are filled with money from many large corporations and belong more or less to the employees. In order to guarantee an increase of the value of the funds, US fund managers decided in the 1970s also to invest in startups (Gompers & Lerner 2004, p.9). With this, not only the pure volume but also the awareness for VC had been strongly increased in the US. Germany could also follow this path. Thus, more VC money could come from the profits of the large corporations as well as from pension funds investing in venture capital.

If a 1% investment into venture funds had been obligatory for just the top 100 companies of Germany in 2010, it would have led to more than 600m EUR extra VC money, due to a total profit of the these companies of over 60bn EUR (Süddeutsche Zeitung 2011; FAZ 2013; Wikipedia 2011). That possible obligation would have meant a doubling of the VC volume in Germany. The same amount could have already been reached in 2012 with the sales of the top 20 companies, which accounted for more than 60bn EUR in profits (FAZ 2013). And a year earlier in 2011, the 30 DAX companies accounted for over 100bn in profits that could have created 1bn EUR extra venture money with the 1% rule (de.statista 2012). As explained in chapter 6.8, the interaction and relation among the elements will have a chain reaction within the ecosystem. Doubling the venture money flow in Germany, would lead to more investments and more investments in different phases. This could attract more foreign top entrepreneurs and employees to found a startup, to finance or to work for a startup in Germany. More top performing foreigners and more money already in the ecosystem will also attract more foreign investors to look for potential targets in Germany. More foreign input, expertise and resources will make it easier for German startups to operate also on a larger European scale. This is because several languages or even English could become the company language, and the foreign EU co-founders as well as employees could provide better access and knowledge to and on other European countries. Foreign entrepreneurs and employees that leave their mother country to work in Germany will also bring new cultural aspects into the ecosystem. They can especially provoke the German co-founders and colleagues, and dare them to try out something new or different, because that is what the foreigners have been doing themselves after having left their country.

7.3.2   Special Tax Treatment

After a nation becomes aware of the importance of the startups for its economy, it needs to further them specially. This is particularly the case, when taxes are involved. Especially in the early years most startups do not have the money for a tax consultant or expertise as well as resources to handle multiple tax issues. Many young companies are very fragile and have almost no financial assets. When they make their first profits, it does not necessarily mean they are strong and mature enough for future growth and stability. Letting startups pay taxes in its first years is very dangerous: On the hand, the taxes reduce the financial means for upcoming activities, on the other hand, the complexity and workload increases through the tax regulations, taking away important time, which should rather be used for more relevant topics. Therefore, tax free periods and special tax deductions for growth should be discussed.

Also the tax obligations of the investors are affected by tax regulations. In 2012 and 2013 the German Business Angel Association (BAND) on behalf of many investors was lobbying for the preservation of certain investor's rights. It wanted to secure tax-free capital gains from startup equity, in case where the profits are reinvested (BAND 2013). The agreement between the lobby groups and the government, at least for the time being, was to maintain this tax-relief for divested startup equity, which was reinvested into venture financing. Dividends, however, could be taxed. In Britain for example, investors can even deduct 30% of their investment of up to 1m GBP from their personal income tax (HM Revenue & Customs 2013), thus making startup investments more attractive. Especially in Germany, where high tax rates are high and a general more cautious behavior of investors exist, offering such tax reliefs, as in Britain, would certainly help creating more German BAs as well as increase their investing budgets. This could be seen as an effective form of reallocation of money, with money going directly into startup investment and not being filtered through government tax machines. The advantage is that there are already individuals behind the investment taking care of it, and no time and money lost after having been collected through taxes, due to a potential government distribution in this particular field.

7.3.2.1 Tax Free Period

The entrepreneurship scene discusses tax issues vividly. The scene consist of a mixture of startups, entrepreneurs, business angels, investors, incubators, journalists, supporters, industry networks and academics (University of Portland n.d.). In Germany, Prof. Faitin (2008) from the Freie Universität Berlin for example argues that startups should be released from all taxes and even governmental paperwork for several years, till their business can afford to pay taxes and handle the organizational burdens. Often, governments tend to be cautious or reluctant for such tax reliefs, because it opens doors for creative fraud. As a result this fear leads, at least in Germany, to a series of exceptions and requirements, which in the end do not relieve the business of the tax burdens and connected problems.

A solution could be a 2 year tax-free time for startups, with revenues below 100k EUR and profits below 50k EUR. The German government scholarship program "EXIST Gründerstipendium" grants up to three founders a monthly salary of 2k EUR plus expenses of about 20k EUR for a year, a total financial support of approximately 90k EUR. This calculated base demand for a founder and could, therefore, be a good comparison for such a tax-free period and level of all startups. It indicates that the government is realizes that a startup needs to earn around 100k EUR a year, in order to survive. Any other tax burdens at this stage could be counterproductive. Even if the startup manages a profit in these first years, within the 100k EUR range, it should have government support to invest the money for future growth rather than paying taxes. The 100k EUR free tax zone would make considerations about paying taxes and hiring a tax consultant obsolete, and thus, at this stage, save resources for the essential goals like growth, market penetration and profitability. In Germany, this kind of support must to be transparent and limited, due to a fear that advantages for a certain group are unfair and can be abused. Limits could be

- the time, for example 2 years after the founding of a company,
- the revenue and or profit size, for example 100k EUR,
- the number of employees and founders, maybe being less than 5 persons, and
- finally some scrutiny of persons that are using this free-tax time more than once. A simplified document could be required, in order to full fill the mentioned considerations above, without creating new tax complexity, which is as high as the current corporate taxation process and documentation.

7.3.2.2 Tax Deductions for Employment Growth

Before paying taxes, at the early stage of a startup life, one has to keep in mind that besides finding new customers the biggest challenge for a startup is to find, hire and pay its first employees. Even after having made a profit within one year, young companies are very fragile about the future of their business, hardly having any track record, limited experience or few sincere customer relations. Hiring a new employee, based upon the growth and the profit from one year, is very hard. By this time, the government already has or expects corporate taxes from the company; money, which could rather be used or saved for the next year's payments for new employees. The author, therefore, recommends a tax-free break, if in fact new employees are hired in the following year. In Germany, at least for the investment of machinery or similar items, this is considered as a pre-depreciations for future investments. However, this is not a tax relief, nevertheless it stills help a company to be more flexible when dealing with its profits and investments. Yet it also shows that Germany is still too focused on its established industries like automotive, chemistry and industrial engineering. It ignores the basic fact and reality that almost 70% of its industry is a service industry requiring different means (Statistisches Bundesamt 2013).

In order to create more jobs and give startups, as well as other types of young companies, a push for further growth, they should also receive a tax relief on profits when creating jobs. The limit could be a 500k EUR turnover and a profit of 100k EUR. These 100k EUR leave space for hiring approximately one to three new employees, with a monthly salary of 6.5k EUR for one experienced sales person or developer, or a base salary at around 2.5k EUR for 3 young professionals.

7.3.3   Opening the Labor Market for More Talents

The German labor market – like most in the world – needs to be more open for talents from around the world. The leaders and founders of top performing US companies supposedly, in 40% of the cases, have a non-US background (Forbes 2014a). In Germany for example, a company not only has to show that a foreign employee is necessary, but also no German employee could do the job. Furthermore, there is also a minimum wage for non-EU employees, which is quite high and leaves little room for flexibility. The German system does not consider variable performance based wages and demands a minimum of 2.5k EUR salary per month, which is more than 15€ per hour on a classic 40 hour per week contract (Arbeitgeberservice 2010). Even though those levels appear to be quite low for established companies, but for startups these are considerable high monthly expenses. The companies need to have more flexibility, in order to reduce the initial hurdle, in order to start hiring employees. Allowing a variable performance based wage level, for startups hiring foreign employees, is one solution, which does not undermine the existing laws, which often forget that usually startups have no secure revenue stream or stable customer relation, which could guarantee a relative secure financial income stream. Since a flexible sales or employee force is necessary to build up the business, a pro-startup governmental framework certainly would increase the chances for those young companies to grow. Since a resident of the European Union is allowed to work and live in each member state, without a permission of the local authorities, extending the laws also for non-EU employees would also be able to attract more specialists from emerging countries like India, China or Russia. This extension of the law for qualified non-EU employees would also help in creating a more united EU market, which is important for the startup and the general business scene. The reasons for this will be further explained furthermore below

7.3.4   New Bankruptcy Law for Startups

In chapter 4, the bankruptcy laws of Germany and the US were compared. Although the author believes that an entrepreneur will found a company despite a good or bad insolvency regulation, there certainly could be a positive influence on the startup scene with the creation better laws in this area. The influence will probably arise from the on going cultural change, which will have with a long-term effect. When reducing the time and burden of bankruptcy like in the US system, the chances for entrepreneurs to try several, even new and unproven approaches for founding a successful company rises. The new progressive laws may change the current attitude and recommendation of parents and friends, to first gather experience working for a company. Founders may follow their dream immediately in starting their own company, especially when a failure becomes more acceptable and is not punishing the entrepreneur too hard, thus is giving him or her another chance within a relatively short period of time. The concrete law suggestion would be, to relief the founder from his burdens after a year, regardless if he or she plans to found a new company or starts working as an employee. This law probably needs to be aligned with a new reform for banks loaning money to startups and their founders. In Germany, banks like to ask for a joint and several liabilities even for small loans in the lower thousands. This means that founders have to become liable for their companies' debts with all their fortune and assets. On the one hand, many young entrepreneurs do not even have high-valued assets to be valued, thus this requirement of the bank inhibits many founders, borrowing money in the first place. Whereas American seem to have it easier to borrow money from a bank or to receive money through the usage of several credit cards (Focus 2013b; Brand eins 2001) – either for business, housing or college. In contrast, many Germans are rather proud of not having loans and even look disrespectful upon the Americans credit card culture. The combination of strict and long-term bankruptcy laws, as well as requirements for bank loans has a strong influence on the founding culture in Germany. Nevertheless, within Europe, Germany is among the top borrowers for money, although it is the most stable and successful big country in Europe. The mentality of Germans is to borrow only money if they can afford it. People with good incomes start using cheap credit lines on consumption, investments for solar panels as well as housing projects. German government ought to urge banks, however, to give more and easier credits to startups, without excessive and cumbersome personal liability. A possible change in attitude of the banks requires building up more expertise about startups and evaluating their concepts and businesses possibilities, rather than giving the credit depending on the personal assets of individuals. At least the US banks seem to have the right attitude and willingness to take the risk. The KfW Bank in Germany (a public federal bank) seems to change its attitude, by giving bank loans to startups under much better conditions and in combination with taking over liability through the *Bürgschaftsbank* ('public liability bank'). However, the process is handled by the local bank, where each startup is located, thus bringing up the same problems as mentioned in the section of debt financing 3.10.4, because parts of the liability remains with the founder and the local bank, thus urging the local bank to apply its general guidelines.

Integrating former or active entrepreneurs with a successful track record certainly could help the evaluation process. Banks need to get in touch and pay, however, the entrepreneurs on a project base, in order to receive their advice and services.

7.3.5   Laws Protecting Startups against Written Warnings

A major problem for startups and small companies in Germany are the written warnings by individuals or law offices, who specialize on finding legal mistakes on the companies' websites. Although they are not supported or enforced by a federal institution, they are establishing their claims on certain German laws. The original idea of these laws was, to protect the consumer from misleading information and criminal acts online. However, these persons or lawyers are taking advantage of this law for their own gain, thus the protective idea discovering such innocent mistakes or violations could lead to law suits and several thousand Euros worth of damage for the new company (Henn 2008)

Many startup teams in Germany have much respect for and fear of these written warnings. The startups cannot afford legal counsel and thus often hesitate to move forward without legal services. Instead founding their company faster, many entrepreneurs spend extra hours on reading the profiles and texts of other websites to understand, how to avoid those written warnings. This experience is reported in many network startup meetings and can be heard in shared startup offices around the country – from Berlin to Munich. Both, the venture managers and the startups do not see any benefits of these warnings and the need for avoiding them. Neither do the startups prepare themselves better for the market entry through this, nor do they learn more about the market and its customers. This effort and fear is always there, till the company can afford to pay regular legal advice and protection.

The time and money spent should rather be used to bring the startups forward and not limit or slow them down. The law should differentiate between simple, unintentional mistakes and obvious intentional fraud intention. In addition, a law could forbid individuals or companies to make money on these claims.

7.3.6   Gender Support

Women make up for only 1/3 of founders and less than 1/5 of startup founders (Chapter 5.10). This ratio is similar in many other countries, and thus these countries think of incentives for women to become more involved in entrepreneurship (Global Entrepreneurship Monitor 2012). However, as pointed out in section 6.4, certain skills and special characteristics of an entrepreneur cannot be acquired. Therefore, one can raise the question, whether it is effective and efficient to focus on one general group like women, instead of looking for mixed gender groups that have already shown a sufficient interest for entrepreneurship. At the moment, it appears based upon current statistics that women are even naturally less inclined to found a company. This fact correlates with the results of the study about women being less competitive than men. Instead of stimulating women to become entrepreneurs, it would be more fruitful to sweep aside hurdles for women, who are already inclined to be entrepreneurs. Through various interviews with female students at the University of Karlsruhe, as well as observed in studies from the WZB in Berlin (Allmendinger & Haarbrücker 2013), family planning still plays a major role in a women's career planning. And since the time frame, in which women want to have children, lies in the same age range as most founders tend to be – 25-44 years (KfW-Research 2012, p.77; Allmendinger & Haarbrücker 2013, p.33ff) – this reality could be seen as a major hurdle for women to found a company. Programs and incentives need to be built around this reality. A general initiative providing *Kitas* ('day care facilities for infants and small children') for all under 3 year old kids has been fostered by the government over recent years, thus women (in parts also men) in general can start working earlier again. Over half a million kids under the age of 3 have been going to *Kitas* in 2012, which stands for 27,6% of this group (Federal Ministry of Family Affaires 2013). Although according to the Federal Ministry of Family Affaires, there were already sufficient *Kitas* places in 2013 with day care supply and demand being around 800K, it nevertheless is difficult for families to find a accessible and especially an inexpensive *Kita* for their child in major cities (Focus 2013, p.30ff). And startups are mainly being build up in those major cities, because of the better infrastructure, universities, established corporate environment, more investors, qualified employees etc. (Die Zeit 2014). A priority for female founders in receiving a Kita place and having it subsidized could take away this hurdle for women to start their own company or be part of a startup foundation.

Another consideration, which could have a major influence on women founding a startup, centers around the state prescribed payment women are to receive during their first year of their child's life in relation to their last income. Per law, they are to receive 67% of the average net income, but no more than a maximum of 1,800€ per month. Thus, the fixed income that a women receives from an employer before giving birth to child provides her more money and security than having a startup, which probably was unable to pay her a salary at all (BMfSFJ 2010). An amendment to this legislation could help here. It could help women that have founded a company should to receive a fixed amount out of this public funding, regardless of what their startup salary was before their pregnancy. Since Germany is very hesitant when it comes to law changes that open the door to potential fraud, the framework of when a pregnant woman, who founded a company, needs to receive these subsidies, requires to be clearly defined. First of all, the female founder needs to show her intentions and the realistic chances of continuing the business after having received government grants for one year. Second, "proof points" of the company's sincerity should be shown, such as:

- The number of initial sales
- A list of customers
- The number of co-founders
- A profound business plan
- Proof of collaboration with other companies or institutions for the development of the product
- Reception of additional funds, grants and awards

All these and more are good signs for sincerity and serious business intentions.

7.4   One European Startup and Venture Market

The challenge of the German startup and venture market is to overcome the limited and much smaller markets in comparison to the USA. Taking the results from the renowned RedHerring's Global Top 100 startups list, Germany had only 2 companies in 2012 and 3 companies in 2011 & 2013 listed, while the US had 48 in 2011, 46 in 2012 and 23 in 2013 listed. Europe together, however, had 30 listings in 2011, 22 in 2012 and 21 in 2013, showing how much stronger Europe would be in a more united venture and entrepreneurial market (RedHerring 2011; RedHerring 2012; RedHerring 2013) – the complete list can be found in the appendix. A more united and homogenous EU market could be an answer for it.

7.4.1   The Venture Capital Perspective

The stated reasons for the US having many more VC funds are multifaceted and among them are the much higher exit opportunities for investors in the US. Large amount of VC money is required in Germany, and elsewhere, in order to be able to reach a critical mass of fund volume (compare 3.11.4 & 3.11.5) and to be able to attract top fund managers. Since numerous ventures are expected to fail or under-perform, a space and a market for at least "3-digit million dollar exit" must be available – either in form of a purchase by another company or by an IPO. This is very rare in domestic markets like Germany, or other European states, simple because the market alone is too small for such exits, and Germany and the other EU countries are currently not seen as the hubs for international growth.

This dilemma could be solved by further uniting the European market, in order to create one huge venture and business market. The former CEO of Lufthansa, Wolfgang Mayrhuber, once said that he considers Europe as the home market and the world as the growth market. This idea should be transformed for the European startup scene. With a EU domestic market of over 500m people, the exit channels will increase, and thus it is more realistic to start several new funds with volumes over 100m. The managers of the funds could of course also come from different European countries, bring in different expertise and thereby break the vicious circle described above. The solution will not be very easy. The Euro, as the common currency, has to be in place in all participating countries, helping to avoid any currency risks and setting the ground for a common market. In addition, tax and investment regulations need to be aligned, in order to reduce the deal complexity. One of the main problems, however, are the different languages used in Europe, which not only make the deal more complex for the investors, but especially stand for a great hurdle for startup to grow in Europe.

7.4.2   The Startup Perspective

Aside from the language, the question will be, how startups can psychologically and practically approach these currently different markets, with different cultures, market needs and regulations. The EU and some of its major nations could further EU-wide active startups and make it easier for them to operate on a European level. A first cross border start is the Erasmus for Entrepreneurs program (European Union 2013), which finances young entrepreneurs an internship abroad, with the idea to learn from other entrepreneurs in a foreign country and market. In addition, there should be further exchange between European incubators and small funds, to foster cross border teaming and knowledge transfer. Eventually, the laws and regulations need to be aligned, thus selling a product in Germany should not be more difficult or different from selling it in France or Spain. Furthermore, cross-border regulations for an easy hiring process of interns, employees and freelancers are required, in order to make use of the potential of European talents and price advantages. Working with EU internationals is one of the best and most effective form of creating a common EU startup – and thus also a venture market.

Altogether, this is a Herculean task, which requires the EU states first of all to understand the necessity of this common market and second to take then actions, in order to prepare the ground for it. Along with the other advantages of a common and more united EU market, this could help the EU decision makers to continue this difficult and strenuous ongoing process. The EU has already faced some of these challenges. Under the Danish EU presidency, a proposal in 2012 was made, allowing venture capital and social entrepreneurship funds to obtain an "EU passport", if their total assets are below 500m EUR. These funds can be used in European countries without meeting additional national requirements, if they fulfill some general criteria (Danish Presidency of the Councel of the EU2012 2012).

7.5   Open Markets for International Investors outside the EU

The idea of one European startup and venture market is already partly supported by the EU research program Horizon 2020, which fosters entrepreneurship across Europe with billions of Euros invested in all kinds of business-, educational and institutional areas (European Commission n.d.). Especially Germany, as the European economic engine, should specifically work more on attracting more foreign investors. Of course, this potential of one European startup and venture market needs to be further spread among rich individuals, funds and corporations from emerging countries such as the BRIC.

7.5.1   The German Market as a Safe Haven for Capital

Many wealthy Chinese and Russians have already invested in German real estate and stocks. They consider Germany a safe haven for their money and are currently still getting fair evaluations in comparison to other countries like France or England (German Trade & Invest (GTAI) 2013). Germany could use this positive economic image to attract these persons also for investing in its startups. The major funds in Germany, like Wellington or Earlybird, already collect much of their money abroad from the Arabic countries, like Saudi Arabia or the United Emirates. In order to attract, however, more foreign investors, Germany needs to level more the ground of taxes and restrictions, in order to make it easier for international venture capital to open VC offices in Germany. Better professional structures, including friendly tax regulations and a high variety of funds, in which foreigners could invest, would help and make up part of this foundation. Agencies, such as the German Trade & Invest, could promote these opportunities worldwide. Programs like the doubling of investments by the KfW or the 20% investor's add-on by the government are also attractive possibilities that foreigners need to be aware of. The idea is certainly embedded in a comprehensive push and improvement of the whole Entrepreneurial Ecosystem in Germany. Once more startups are financed and eventually become visible and successful, more money will flow in and also attract investors from a far. With the financial capital of the new investors, German startups will also profit, thanks to their connections and knowhow. It is like an avalanche that needs to be set in motion, by one step after the other of the above.

7.5.2   Good Competition for German Investors

In time, German investors will be forced by the increased competition to widen their investment scope. An interesting consequence thereof could be that German investors must also look for in investing in startups abroad, like US investors have done. They are spreading around the world nowadays to identify the best global startups and technologies. Today, unfortunately Germany's top investors are not yet forced to invest in foreign teams or technology, and thus taking an extra risk of unknown markets, because they have enough opportunities with the low competition in Germany. In contrast, American investors are quite interested in German teams and technologies, although many still require an intention to move parts or the whole business to the US. Since US investors also prefer to have a certain proximity to their teams, they began opening investment offices around the globe. Germany needs to encourage its best funds to do the same. However, it must delicately be done, in order not draw too much capital out of the underfinanced German VC market. A close collaboration with the biggest VC funds in Germany is necessary, to understand their needs on the one hand, while increasing the competition on the other hand. The investment in foreign teams can also have a positive influence on the German market.

The German VCs will increase their expertise and might also attract the teams to settle in Germany or at least have a subsidiary there. The mixture of international investors and international teams will finally nurture the German ecosystem. It will thereby also help attracting more international talent to come to Germany. An ecosystem lives from the interaction of its elements. And thus the Entrepreneurial Ecosystem will also respond to improvements on the financial side on other areas.

7.6   Open Culture for Entrepreneurship

The German society needs to become more open for entrepreneurship in general. Since cultural issues cannot simply be changed with a government program, a whole set of changes and initiatives needs to move society into the right direction. This may, however, take a generation till it implements programs and achieves the necessary results, since cultural changes occur mostly in countries, in which the younger generation has a different mindset, experience and different formative conditions – than their parent generation (Inglehart & Welzel 2005, p.113).

Among many others, the following elements play a major role in this cultural change:

- Media coverage
- School & university teaching material
- Special government programs
- New attitude and code of business of corporation
- General people's attitude

Improving the cultural environment, therefore, requires changes and improvement in the Educational and Economy & Business Circle. This is another indicator, why a holistic approach and ecosystem is needed, because the elements really influence each other throughout the different circles. Their interplay creates new aspects, like an entrepreneurial culture, which cannot be simply summoned and explained by the participating elements and their original purpose. It is also impossible to rank the influencing factors on the entrepreneurial culture, since they can only be seen in light of their individual connections.

7.7   Pro Entrepreneurship Media Involvement

The media can help to create and transport a positive image of entrepreneurship as a whole and in general of the main character, the entrepreneur. From talk shows to soap operas, there needs to be a presence of enthusiastic and decent entrepreneurs that stand for a positive company culture, instead of nurturing old negative clichés for high audience rates. The media must be made aware of and has to understand its responsibility, as well think of ways how to attract their viewers with positive new images of businessmen and entrepreneurs. Of course the media by all means should not forget about illegal behavior and fraud of businesspersons as well as companies. It is rather an appeal to the mass media for acknowledging or accepting responsibility, in order to influence positively the development of our economy. Positive developments require public presentation of optimistic examples, role models and show cases. In the long run this might also have an effect for the better on the envy culture of Germany.

7.8   Overview on Potential Changes

In order to have more successful startups in Germany or in a nation with a similar disposition, many elements of the Entrepreneurial Ecosystem need to be improved. By only increasing the amount of VC money, for example, the ecosystem will not necessarily be changed enough, in order to make the startups more successful. Nevertheless, it is one of the key aspects for Germany. Money can be compared with fuel for a car engine. Without fuel, there will be no movement, on the other hand, without a car, its engine or the street to drive on, the fuel is unnecessary and of no importance. The road could be the economy the startup will be placing its business onto, whereas the engine might be the team, pushing the idea forward. The car then would stand for the essential parts of a company like concept, technology and staff. Narrowing founding companies down to such a simplified image can help to understand the connections better, in order to structure and to get started with the improvement of the Entrepreneurial Ecosystem.

The following table shows key improvements, their influences, the timeframe, as well as the activities that need to be done and fulfilled by the key players. Most improvement initiatives are coming from the Public & Business Circle, with some also being connected to the third, the educational circle. Elements from the inner two circles can hardly be changed or influenced directly, however measurements in the outer Circles III and IV should have an indirect effect. In the long run, the experience and skills of potential (serial)-entrepreneurs will also be improved, if they have the chance to start over or start with another project or startup.

Name	Means	Influence	Activities	Time	Key Player
Entrepreneurs in the educational system [Circle II+III]	- Courses, seminars and speeches by entrepreneurs - Entrepreneurs becoming coaches	- More role models - First contacts - Better understanding - Personal adviser	- Inviting successful entrepreneurs from the alumni network and the region - More incentives like academic merits or access to top students	Short	- Universities - Regional networks - Entrepreneurs
Flexible schedules [Circle III]	- Sabbatical year for entrepreneurship - Exam-flexibility - Part-time jobs at the university - Encourage corporations to accept students studying longer	- Time during the studies to get to know startups, without having the pressure right after graduation to decide between employment and entrepreneurship - Financial and social backup with a part-time job ideally related to a potential startup topic	- Stretched time tables - Further education to practice concept aligning demand for employees from the industry with demand for entrepreneurs - Part-time positions - Support statements by corporations to be in favor of longer studying periods for entrepreneurship activity	Med to Long	- Government - Universities - Industry
New Business Planning (BP) [Circle III+IV]	- New guideline and templates - New form of teaching	- Supporting the founding flow - Covering the right fields at the right time - Professionalizing	- Design of new guidelines and templates - Circulation - Adaptation	Short	- Investors - Entrepreneurship institutes - BP Competitions
Business Cases / Scenario Planning [Circle III+IV]	- New guideline and templates - New form of teaching	- Understanding the business better - Matching a BP - Professionalizing	- Design of new guidelines and templates - Circulation - Adaptation	Short	- Investors - Universities - BP competitions
Comprehensive founding material [Circle III+IV]	- New website	- Supporting a quick founding process - Focus on relevant topics	- Central website will all info - Briefing of all official institutions - Concept for continuous content management	Short	- Institutions - Government - Universities

				Time	Actors
Corporate Venture Law (Corporate Profits going to VC funds) [Circle IV]	- New laws and additions - Vehicles for accelerated fund building	- Increasing the available VC amount - Turning the market into a founder's market - More founded startups	- Building of an expert panel - Design phase - Negotiations with corporations - Legislative initiative - Founding of new funds	Med	- Government - Expert panel
Tax Free Period [Circle IV]	- New laws and extensions	- Faster growing startups - More time for the essential business parts	- Building of an expert panel - Design phase - Legislative initiative	Med	- Government
Tax free Reinvestment on employee growth [Circle IV]	- New laws and modifications	- Faster growing startups - Supporting growth	- Building of an expert panel - Design phase - Legislative initiative	Med	- Government
Opening for international talents & teams [Circle IV]	- New law additions - Public campaign - Scholarships	- Better teams - Faster internationalization - Cultural change through international teams	- Establishing an expert panel - Design phase - Legislative initiative	Med	- Government - Universities - Corporations - Society
New Bankruptcy law [Circle IV]	- New laws, additions and revisions	- More startups - More serial entrepreneurs - Potential cultural change	- Establishing an expert panel - Design - Legislative initiative	Med	- Government
Written Warning protection [Circle IV]	- New laws and additions	- Faster startup launches - More time for the essential business parts - Saving money	- Setting up an expert panel - Design phase - Legislative initiative	Med	- Government
Gender support	- Kita Priority - Fixed grants in the first	- More female entrepreneurs	- Amendment of federal laws and repealing or changing	Short to	- Government o Federal

				Med	o State
[Circle IV]	year after given birth to a child				
One EU VC Market [Circle IV]	- EU initiative - New and additional laws and regulations	- More money - More growth potential - Higher evaluations and higher investment rounds - More key IT players in the EU	- Building of a German expert panel - Design phase - Building of EU expert panel - Design continuation - EU processes	Long	- European governments - VC funds
More ex EU investors [Circle IV]	- New and additional laws - Public campaign - PPP fund for non EU investors	- More money - More funded startups, increasing variety - Higher evaluations and higher investment rounds	- Attractive tax laws - Building up a whole set of new funds - International marketing campaigns	Med	- Government - New funds by Business Angels and entrepreneurs
Open Culture [No direct connecting, rather Circle IV]	- Support entrepreneurship on various other levels	- General improvement on all circles of the Entrepr. Ecosystem and elements - Catalyst for further change - Guaranteeing sustainability	- Media campaigns - Reform of educational material - Role models stepping up - Implementing the laws above - Inviting more foreigners	Long	- Society - Corporations - Government
Media coverage [Circle IV]	- Media campaigns encouraging more entrepreneurship	- Cultural change through opening society for entrepreneurship	- Conference by all major newspapers/magazines to further the understanding of their responsibility and to identify helpful measurements - Creating new TV formats and series for "good" business	Med to Long	- Mass media

Table 38: Entrepreneurial Ecosystem Improvements

8 Summary & Outlook

"Mir hilft der Geist! Auf einmal seh ich Rath und schreibe getrost: Im Anfang war die Tath"[8]. This famous quote by Goethe can also be applied to entrepreneurship. After all, entrepreneurship requires doing. The best ideas, smartest persons or most developed ecosystems are only helpful, if the entrepreneur takes action. This certainly is easier said than done, in light of all the insecurity and challenges that are lying ahead of the entrepreneurs and their startups. However, if the entrepreneur takes action, the Entrepreneurial Ecosystem can be decisive for his or her success.

The author has shown that the Entrepreneurial Ecosystem is being mentioned imprecisely as a term in many academic and business sources around the world. Yet, it can be defined much more accurately, and through this new more holistic definition, improvement points can be identified and thus implemented. In order to develop this comprehensive definition and in order to find improvement points, literature research, surveys among investors and entrepreneurs, as well as a comparison between the US – the leading entrepreneurial nation of the world – and Germany – the economic engine of Europe – in terms of their entrepreneurial market aspects have been conducted.

The author's holistic concept consists of 4 circles with a total of 24 elements that are connected and influence each other. In the center is the Personal Circle of the entrepreneur with his or her key elements. All other circles arise around the entrepreneur and its Personal Circle and stand in relation to the four main elements in this circle.

This ecosystem is a concept that can be used as a template to investigate and to compare the entrepreneurial situation of each country. The concept has been created in such a way that it serves all types of innovation and entrepreneurship activity, whether coming from Academia or from persons, without any academic background. The concept can be adapted, and elements can be further added or replaced, depending on the country's circumstances. A quantitative benchmark, however, with a valuation and prioritization of the elements, in order to create an index, which can be measured and compared over time, lies beyond the scope of this thesis. These considerations could be the research field of another study.

Improvement suggestions in this ecosystem can primarily be found and developed in the two outer circles, the Educational Circle (Circle III) and the Public & Business Circle (Circle IV). For example in Circle IV, government, media and the public itself should support and cheer these founders, increasing their standing and importance for society.

---

[8] "The spirit is helping me, and I am seeing advice and thus I can write: In the beginning there was the action"

Opening markets and truly living internationalization will not only help a nation to have a successful Entrepreneurial Ecosystem, but also increase chances for prosperity and peace across other nations. With the examples of the many successful startups from the US that have been imitated or copied by German companies, one can understand the importance of market sizes, growth and exit potential much better. It should have become obvious, how the entrepreneurial scene is driven by internationalization, and thus is providing potential for Germany to learn from the US ecosystem. Europe is providing the foundation and size for creating the required international stage. Some entrepreneurs and investors are already playing on this European field, and the aim of this thesis was to further encourage the German stakeholders, especially the government, to quickly nurture, encourage and additionally stimulate this business development.

From the startup perspective, a starting element for an avalanche of events for the improvement of the Entrepreneurial Ecosystem is risk financing. The author's market analysis has shown that Germany clearly lacks in this field behind the US, as well as with its entrepreneurial culture and needs to undertake a number of steps to close the gap, in order to become a more successful entrepreneurial nation in the 21st Century. Germany must change the investor's market into a founder's market, in order to have an increasing and more successful startup scene. Engaging large corporations to invest parts of their profits into venture capital is one option. Many improvement points can and will follow. Again, it is also the perspective of a united Europe that bears the potential to change this financial under-financing.

A positive influence on the entrepreneur from the Private Circle, with certain individuals such as family and friends, has also been shown through this paper – from role modeling to financial support.

Further improvement ideas have been shown in the outer circle, such as the author's independently developed "golf-layout" business plan. With this business plan, the author has tried to provide an alternative and an escape from the chronological dilemma of commonly used business plans. This setup now needs to be used, tested and altered to have a positive effect on a startup's business.

All the influences mentioned in this thesis shall increase the success chances of an entrepreneur. For the author it is important, that this success is not only related to pure financial results, which is also the case nowadays for many other individuals. Social entrepreneurship, increasing donations from rich entrepreneurs, as well as the general attitude one can feel among young entrepreneurs, support this idea that success consists of more than just making or acquiring money. Nevertheless, one should not make the mistake thinking, money is not the major motivation for founding and running a company. Money clearly dominates this domain, from the financing to the establishment and "exit" of the company.

Success will foster innovation and is essential for each nation. Entrepreneurship is an important vehicle, which brings innovation to life. The thesis outlined other forms of innovation, from a pure corporate to an intra- and entrepreneurial innovation. The focus of the thesis however, has been on the entrepreneur side with the new Entrepreneurial Ecosystem.

Considering the positive all-around results of modern innovative companies, one can see many commercial, medical and other improvements for our lives – from search engines, to car sharing, from smartphones to medical treatments. Many examples could be found to show, how innovation is improving our lives by making things better, safer and more enjoyable. Aside from these major life relevant challenges, innovation can help to raise the standard of living, by providing better quality, lower prices and less dangerous components.

The thesis also derived three waves of innovation that have started as early as two thousand years ago. And potentially, a fourth wave coming from a social entrepreneurial side is already in motion.

These social entrepreneurs are symbols for thoughtful businesspersons that lead their companies and employees into the future. Entrepreneurs must find a balance between personal and public goals and spread a positive success definition of their own across their ecosystem.

Ultimately, real entrepreneurs will not wait for governments to take action and clear their path. Turbulences, imbalanced markets, technology shifts or increasing population and general wealth always open potential for new ventures. Where fearless entrepreneurs found or build up their companies, create jobs and pay taxes, is the challenge for each nation to recognize and to create an environment to attract them.

Appendix

The Appencix of this thesis can be accessed on www.springer.com under Christian Schwarzkopf in the OnlinePlus-Program.

Appendix A

In Appendix A the complete survey questions and answers from the investor's perspective are listed. The first question asked the participants whether they would like to receive a summary of the survey results. The total number of countable participants was 28. Not all questions were obligatory,, however most of them have been answered by all 28. Questions 2 to 31 with the results are.

Appendix B

In Appendix B the complete survey questions and answers from the startup's perspective are listed. Over 90% of the monetary numbers were given in EUR (question 13), thus a split between EUR and USD has not been conducted for the analysis of the answers. The total number of countable participants was 62. Not all questions were obligatory, thus the average number of answering participants was over 40. The lowest number of answers for a question was 28, the highest 62.

Appendix C

In Appendix C the complete Red Herring List of the Top 100 Word companies between 2013 – 2011 can be seen.

References

1000Ventures.com, 2013. Venture Financing Stages. Available at: http://www.1000ventures.com/venture_financing/venture_funding_stages.html.

About.com, 2014. 15th Century - the technology, science, and inventio. Available at: http://inventors.about.com/od/timelines/a/Fifteenth.htm.

Adler, A., 1996. *Menschenkenntnis*, Fischer.

Airbnb, 2008. Airbnb Pitch Deck. Available at: http://www.businessinsider.com/airbnb-a-13-billion-dollar-startups-first-ever-pitch-deck-2011-9#-1.

Akerlof, G.A., 1970. The market for" lemons": Quality uncertainty and the market mechanism. *The quaterly journal of economics*, (84(3)), pp.488–500.

Albisetti, J., 1983. *Secondary School Reform in Imperial Germany*, Princeton University Press.

Aldrich, H. & Zimmer, C., 1986. Entrepreneurship through social networks. *University of Illinois at Urbana-Champaign's Academy for Entrepreneurial Leadership Historical Research Reference in Entrepreneurship*.

Allmendinger, J. & Haarbrücker, J., 2013. *Lebensentwürfe heute Wie junge Frauen und Männer in Deutschland leben wollen*,

Amabile, T.M., 1996. Assessing the work environment for creativity. *Academy of Management Journal*, (39(5)), pp.1154–1184.

Anderson, D. et al., 2012. *Quantitative Methods for Business*, Cengage Learning.

Anderson, D., 2013. Top 30 Succesfull Online Business Gurus. *David Anderson Wealth*. Available at: http://www.davidandersonwealth.com/entrepreneurs/top-30-successful-online-business-gurus-and-why-they-are-so-effective/.

Anon, 2014. Hard Work vs. Talent. *Vistabrokers*. Available at: http://www.vistabrokers.com/philosophy-evolution/philosophy/hard-work-vs-talent-2014-08-12-1.

Anon, The 50 Most Innovative Companies 2010 - BusinessWeek. Available at: http://www.businessweek.com/interactive_reports/innovative_companies_2010.html [Accessed October 7, 2010].

Arbeitgeberservice, 2010. Labor costs for foreigners.

Arie, G., 2014. US Venture funds investing in Europe from a lawyers perspective.

Aristotles, 350AD. The Nichomean Ethics.

Ashoka, 2013. Def. Social Entrepreneurs. Available at: https://www.ashoka.org/social_ entrepreneur [Accessed January 28, 2013].

Asquith, P. & Mullins, D.W., 1986. Equity issues and offering dilution. *Journal of Financial Economics*, (15(1-2)).

Babinotis, G., The Greek Language as a Basis of Intellectual Creativity. Available at: http://www.babiniotis.gr/wmt/webpages/index.php?lid=2&pid=7&catid=M&apprec =23.

Backes-Gellner, U.P.D., Gerybadze, A.P.D. & Harhoff, D.P.P., 2012. *Gutachten zu Forschung, Innovation und technologischer Leistungsfähigkeit in Deutschland 2012.*, Expertenkommission Forschung und Innovation (EFI). Available at: http://www.stifterverband.de/statistik_und_analysen/wissenschaftsstatistik/efi/efi_ jahresgutachten_2012.pdf.

Bafa, 2013. Wagniskapitalzuschlag. Available at: http://www.bafa.de/bafa/de/wirt schaftsfoerderung/investitionszuschuss_wagniskapital/index.html.

Baker, J.T., 2003. *Andrew Carnegie: Robber Baron As American Hero*, WADSWORTH Incorporated.

BAND, 2013. Steuern auf Veräußerungsgewinne aus Streubesitz? Available at: http://www.business-angels.de/brandneues-122013-steuern-auf-verauserungsgewinne-aus-streubesitz/.

Baskin, J., 1989. An Empirical Investigation of the Pecking Order Hypothesis. *Financial Management*, pp.26–35.

Batchelor, R., 1994. *Henry Ford, mass production, modernism, and design*, Manchester University Press.

Beattie, A., Market Crashes: The Dotcom Crash. *Investopedia*. Available at: http://www.investopedia.com/features/crashes/crashes8.asp.

Berger, A. & Udell, G., 2005. *Small business and debt finance, Handbook of Entrepreneurship*,

Bhide, A., 1994. How entrepreneurs craft. *Harvard Business Review*. Available at: http://serempreendedor.files.wordpress.com/2008/09/how-entrepreneurs-craft.pdf.

Block, J. & Sandner, P., 2009. What is the Effect of the Financial Crisis on Venture Capital Financing?

Bloomberg, 2013. 50 Most innovative countries. *www.bloomberg.com*. Available at: http://www.bloomberg.com/slideshow/2013-02-01/50-most-innovative-countries. html.

Bloomberg, 2012. Banks Seen Dangerous Defying Obama's Too-Big-to-Fail Move. *www.bloomberg.com*. Available at: http://www.bloomberg.com/news/2012-04-16/obama-bid-to-end-too-big-to-fail-undercut-as-banks-grow.html.

Bloomberg, 2014. Google's Nest to Buy Security Startup Dropcam for $555 Million. *www.bloomberg.com*. Available at: http://www.bloomberg.com/news/2014-06-21/google-s-nest-buying-security-company-dropcam-for-555-million.html.

Bloomberg, 2011. TA Associates, Summit Partners Agree to Buy Bigpoint on Online-Gaming Boom. *Bloomberg.com*. Available at: http://www.bloomberg.com/news/ 2011-04-26/ta-associates-summit-partners-agree-to-acquire-bigpoint-on-web-game s-boom.html.

Bloomberg, M., 2013. Bloomberg: Why Sandy forced cities to take lead on climate change. Available at: http://edition.cnn.com/2013/08/21/world/europe/bloomberg -why-sandy-force-cities/.

Blumberg, D., 2015. Berlin based Start-Up Companies - a new attraction for Venture Capital investors from the UK and the US.

BMfSFJ, 2010. Elterngeld und Elternzeit. Available at: http://www.bmfsfj.de/Redaktion BMFSFJ/Broschuerenstelle/Pdf-Anlagen/Elterngeld-Vorlese.PDF,property=pdf,berei ch=bmfsfj,sprache=de,rwb=true.pdf.

BMWi, 2014a. BMWi Business Plan. Available at: http://www.existenzgruender. de/gruendungswerkstatt/businessplaner/.

BMWi, Die wirtschaftspolitische Entwicklung von 1949 bis heute. *Bundesministerium für Wirtschaft und Technologie (BMWi) / Geschichte*. Available at: http://www. bmwi.de/DE/Ministerium/Geschichte/wirtschaftspolitik-seit-1949.html.

BMWi, 2014b. EXIST Programs. Available at: http://www.bmwi.de/DE/Themen/ Technologie/Innovationsfoerderung-Mittelstand/wagniskapital-unternehmensgruen dungen,did=377306.html.

BMWi, 2014c. Unternehmergeist in die Schulen. Available at: http://www.unternehmergeist-macht-schule.de/DE/Initiativen/initiativen_node. html.

boerse.de, 2014. Dax-Titel: Gründerzeit-Unternehmen dominieren. *boers.de*. Available at: http://www.boerse.de/geld/Dax-Titel-Gruenderzeit-Unternehmen-dominieren/ 7463609.

Boldt, K., 2011. StudiVZ-Verkauf gescheitert. *Manager Magazin Online*. Available at: http://www.manager-magazin.de/unternehmen/artikel/a-775584.html.

Bolter, N., 2010. *Coaching for Character: Mechanisms of Influence on Adolescent Athletes" Sportsmanship*. University of Minnesota. Available at: http://conservancy. umn.edu/bitstream/handle/11299/95905/Bolter_umn_0130E_11394.pdf?sequence =1 [Accessed January 5, 2015].

Boocock, G. & Woods, M., 1997. The evaluation criteria used by venture capitalists: evidence from a UK venture fund. *International Small Business Journal*, (16(1)), p.36.

Botazzi, L. & Da Rin, M., 2002. Venture Capital in Europe and the Financing of Innovative Companies. In *Economic Policies*. v34. pp. 229–269.

BPC, 2014. Bizplancompetitions. Available at: http://www.bizplancompetitions.com/.

Brand eins, 2001. Das Kreditkartenhaus. , (07/2001). Available at: http://www.brandeins.de/archiv/2001/geld/das-kreditkartenhaus.html.

Brandis, H. & Whitmire, J., 2011. Turning Venture Capital Data into Wisdom: Why Deal Performance in Europe is now Outpacing the US. Available at: http://www.slideshare.net/earlybirdjason/earlybird-europe-venture-capital-report [Accessed March 19, 2012].

Brandis, H. & Whitmire, J., 2011. *Turning Venture Capital Data into Wisdom: Why Deal Performance in Europe is now Outpacing the US. PDF-Slides*,

Braxton Finance, 2013. The case (against) the 5 year financial projections. Available at: http://braxtonfin.com/blog/2013/6/9/the-case-against-the-5-year-financial-projections.

Brettel, M.P.D., 2002. *Entscheidungskriterien von Venture Capitalists*,

Brugger, R., 2009. *Der IT Business Case*, Springer Berlin Heidelberg.

Bundesministerium für Finanzen ed., 2009. Die wichtigsten Steuern im internationalen Vergleich.

Bundesministerium für Finanzen, 2012. Die wichtigsten Steuern im internationalen Vergleich.

Bundesministerium für Justiz, 2014. *Beteiligung an anderen Körperschaften und Personenvereinigungen*, Available at: http://www.gesetze-im-internet.de/kstg_1977/__8b.html [Accessed January 13, 2015].

Bundesministeriums der Justiz und für Verbraucherschutz, 2009. Einkommensteuergesetz (EStG). Available at: http://www.gesetze-im-internet.de/bundesrecht/estg/gesamt.pdf.

Bundesverband Deutscher Kapitalbeteiligungsgesellschaften (BVK), 2013. *BVK's annual statistics for 2012: a strong final spurt for the German equity capital market*, Available at: http://www.bvkap.de/privateequity.php/cat/67/aid/814/title/BVK%27s_annual_statistics_for_2012:_a_strong_final_spurt_for_the_German_equity_capital_market.

Bundesverband Deutscher Kapitalbeteiligungsgesellschaften (BVK), 2011a. *Das Jahr 2010 in Zahlen*, Available at: http://www.bvkap.de/privateequity.php/cat/42/title/Aktuelle_Statistiken.

Bundesverband Deutscher Kapitalbeteiligungsgesellschaften (BVK), 2011b. Das Jahr 2010 in Zahlen. Available at: http://www.bvkap.de/privateequity.php/cat/42/title/Aktuelle_Statistiken [Accessed February 27, 2012].

Burgstone, J & Murphy, B., 2012. *Breakthrough Entrepreneurship*, Farallon Publishing.

Burke, A. & Hanley, A., 2006. Bank Interest Margins And Business Start-Up Collateral: Testing For Convexity. *Scottish Journal of Political Economy*, (52(3)), pp.319–334.

Burton, K.D. 2011. *Managing Emerging Risk*, CRC Press.

Burt, R., Kilduff, M. & Tasselli, S., 2013. *Social Network Analysis: Foundations and Frontiers on Advantage*,

Business Insider, 2011. The Key Differences In VC Financing Of IT Startups In The U.S., UK, Germany And France. Available at: http://www.businessinsider.com/differences-in-venture-capital-financing-of-us-uk-german-and-french-information-technology-startups-2011-5.

Business Insider, 2013. The Largest Ancestry Groups In The United States. *www.businessinsider.com*. Available at: http://www.businessinsider.com/largest-ethnic-groups-in-america-2013-8.

Business Net Partners, 2013. Business Plan & Case for an International Machine Building Company.

Businessweek, 2007. How MP3 Was Born. *businessweek.com*. Available at: http://www.businessweek.com/stories/2007-03-05/how-mp3-was-bornbusinessweek-business-news-stock-market-and-financial-advice.

BVCA, 2002. Limited Partnership Agreement. Available at: http://www.bvca.co.uk/ Portals/0/library/Files/StandardIndustryDocuments/LPAgreement.pdf.

Byers, T., Kist, H. & Sutton, R.I., 1997. Characteristics of the Entrepreneur: Social Creatures, Not Solo Heroes. In *The Handbook of Technology Managament*. Available at: http://www.stanford.edu/class/e140/e140a/content/Characteristics.html [Accessed March 28, 2012].

Cannon, J. & Perreault, W., 1999. Buyer-Seller Relationships in Business Markets. *Journal of Marketing Research*, 36, pp.439–460.

Caprino, K., 2013. How Different Are Female And Male Entrepreneurs? Susan Sobbott, President Of American Express OPEN, Weighs In. *Forbes - Leadership*. Available at: http://www.forbes.com/sites/kathycaprino/2013/08/21/how-different-are-female-and-male-entrepreneurs-susan-sobbot-president-of-american-express-open-weighs-in/.

Cassar, G., 2004. The financing of business start-ups. *Journal of Business Venturing*, (19(2)), pp.261–283.

CB Insights, 2012. *Seed Investing Report – Startup Orphans and the Series A Crunch*, Available at: http://www.cbinsights.com/blog/trends/seed-investing-report.

Centre for Strategy & Evaluation Service, 2012. Evaluation of EU Member States' Business Angel Markets and Policies Final report.

Centre for Strategy & Evaluation Service, 2002. Guide to Risk Capital in Regional Policy.

Charities Aid Foundation (CAF), 2013. *World Giving Index 2013*, Charities Aid Foundation. Available at: https://www.cafonline.org/PDF/WorldGivingIndex 2013_1374AWEB.pdf.

Chesbrough, H.W., 2003. *Open Innovation: The New Imperative for Creating and Profiting from Technology*, Harvard Business School Press.

Chorev, S. & Anderson, A.R., 2006. Success in Israeli high-tech start-ups; Critical factors and process. *Technovation*, (26(2)), pp.162–174.

CIE, 2012. *Abschlussbericht Centers für Innovation & Entrepreneurship von 2008-2011*,

Cloer, T., 2012. DaWanda erhält Millionen-Investment. *Computerbild*. Available at: http://www.computerwoche.de/a/dawanda-erhaelt-millionen-investment,2509653.

CNN, 2013. OECD: U.S. will recover faster, Europe faces unemployment crisis. *www.cnn.com*. Available at: http://edition.cnn.com/2013/05/29/business/oecd-u-s-europe-economic-recovery/.

Code.org, 2013. *What Most Schools Don't Teach*, Available at: https://www.youtube.com/watch?v=nKIu9yen5nc.

Co, J. & Groenwald, J., 2006. *Fresh Perspectives: Entrepreneurship*, Pearson Education South Africa.

Colpan, M., 2008. Von der Hochschule zum eigenen weltumspannenden Industrieunternehmen.

Compamedia, 2013. Die innovativsten Unternehmen im Mittelstand. *Top 100*. Available at: http://www.top100.de/die-top-100/uebersicht-2013/index.html.

Cooper, A.C., Gimeno-Gascon, F.J. & Woo, C.Y., 1994. Initial human and financial capital as predictors of new venture performance. *Journal of Business Venturing*, (9(5)), pp.371–394.

Cooper, R.G., 1996. Overhauling the new product process. *Industrial Marketing Management*, 25, pp.465–482.

Cooper, R.G. & Kleinschmidt, E.J., 1990. *New products - the key factors in success*, American Marketing Association.

Cormen, T.H. 2001. *Algorithms*, MIT Press.

Countryeconomy, 2014. GDP - Gross Domestic Product. *Countryeconomy.com*. Available at: http://countryeconomy.com/gdp.

Crowdfund Insider, 2014. Crowdfunding in Germany Today. Available at: http://www.crowdfundinsider.com/2014/01/30685-crowdfunding-germany/.

Crunchbase, 2014. Funding Rounds. Available at: http://www.crunchbase.com/funding-rounds?page=1.

Cumming, D. & MacIntosh, J.G., 2003. A cross-country comparison of full and partial venture capital exits. *Journal of Banking & Finance*, (27(3)), pp.511–548.

DAA-Stiftung, Definition Abilities & Skills. *DAA-Stiftung*. Available at: http://www.daa-stiftung.de/begriffserklaerungen.htm [Accessed December 3, 2012].

Daily Mail UK, 2010. 5,000 friends on Facebook? Scientists prove 150 is the most we can cope with. *DailyMail*. Available at: http://www.dailymail.co.uk/news/article-1245684/5-000-friends-Facebook-Scientists-prove-150-cope-with.html.

Danish Presidency of the Councel of the EU2012, 2012. Better access to venture capital in the EU. Available at: http://eu2012.dk/en/NewsList/Juni/Uge-26/Funds.

Dante, A., 13th century. *prose e rime liriche*,

Darlin, D., 2006. The iPod Ecosystem - New York Times. *New York Times*. Available at: http://www.nytimes.com/2006/02/03/technology/03ipod.html?ex=&_r=0 [Accessed January 7, 2015].

Davinci Innovations, 2008. Leonardo Da Vinci Inventions. Available at: http://www.davinci-inventions.com/.

Deans, G., Kroeger, F. & Zeisel, S., 2002. *Winning the Merger Endgame*, McGraw Hill.

Al-Debei, M., Avison, D. & El-Haddadeh, R., 2008. Defining the Business Model in the New World of Digital Business. In In Proceedings of the Americas Conference on Information Systems. Toronto.

DeFilippis, J., Saegert, S. & Lebenthal, T., 2013. *Children in families in communities*, Routledge.

Dellabarca, R., 2002. *Understanding the "opportunity recognition process" in entrepreneurship, and consideration of wheter serial entrepreneurs undertake opportunity recognition better than novice entrepreneurs,*. University of Cambridge.

Deloitte, Jensen, M. & Heesen, M., 2010. 2010 Global Trends in Venture Capital: Outlook for the Future. Available at: http://www.deloitte.com/assets/Dcom-UnitedStates/Local%20Assets/Documents/TMT_us_tmt/us_tmt_VC2010Global%20Trend_160910.pdf.

Department of the Treasury, I.R.S. ed., 2011. Tax Guide 2011 for Indiduals.

Department of the Treasury, I.R.S. ed., 2006. Tax Guide for Corporations.

Department of Treasury (IRS), 2006. Taxguide 2006 for Corporations. *irs.gov*.

Derby, D., 2001. How The Stirrup Changed Our World. Available at: http://www.strangehorizons.com/2001/20010924/stirrup.shtml.

Destatis, 2008. Destatis Vereinigte Staaten von Amerika. Available at: ttp://www.destatis.de/jetspeed/portal/cms/Sites/destatis/Internet/DE/Content/Statistiken/Internationales/InternationaleStatistik/Land/Amerika/VereinigteStaatenvonAmerika,template=renderPrint.psm [Accessed July 1, 2011].

Destatis, 2005. Gewerbesteuerhebesätze 2010. Available at: http://www.destatis.de/jetspeed/portal/cms/Sites/destatis/Internet/DE/Presse/pm/2011/08/PD11_298_735,templateId=renderPrint.psml [Accessed February 16, 2012].

de.statista, 2014. Entwicklung der Zahl der Selbstständigen in freien Berufen in Deutschland von 1992 bis 2013 (in 1.000, jeweils zum 1. 1. des Jahres). *Statista*. Available at: http://de.statista.com/statistik/daten/studie/158665/umfrage/freie-berufe---selbststaendige-seit-1992/.

de.statista, 2012. Operativer Gewinn der DAX-30-Unternehmen im Geschäftsjahr 2011 (in Millionen Euro). Available at: http://de.statista.com/statistik/daten/studie/223592/umfrage/operativer-gewinn-der-dax-unternehmen/.

Deutscher Bundestag, 2013. Bundesrepublik Deutschland. *Bundestag.de*. Available at: http://www.bundestag.de/kulturundgeschichte/geschichte/parlamentarismus/brd_parlamentarismus.

Deutscher Bundestag - Unterausschuss Neue Medien, 2011. *Unterausschuss Neue Medien (22)*, Available at: http://gruen-digital.de/wp-content/uploads/2011/06/2011-05-09_protokoll_existenzgruendung.pdf.

Deutscher Startup Monitor, 2013. Deutscher Startup Monitor 2013.

Deutsches Museum, 2000. *Meisterwerke aus dem Deutschen Museum III*, Available at: http://www.deutsches-museum.de/de/sammlungen/ausgewaehlte-objekte/meisterwerke-iii/dampfmaschine/.

Deutsche-Startups.de, 2012. Investitionsphasen. *ds startups*. Available at: http://www.deutsche-startups.de/lexikon/investitionsphasen/.

Deutsche-Startups.de, 2013a. Top 100 Startups 2012 in Germany. Available at: http://www.deutsche-startups.de/startups-2012/.

Deutsche-Startups.de, 2013b. Zalando: Umsatz schießt weiter in die Höhe; Bewertung steigt auf 3,6 Milliarden Euro. *Deutsche Startups.de*. Available at: http://www.deutsche-startups.de/2013/10/23/zalando-umsatz-bewertung/.

Dien, A.E., 1986. *The Stirrup and Its Effect on Chinese Military History*,

Die Welt, 2012. Pfusch am Bau ist in China an der Tagesordnung. *Die Welt*. Available at: http://www.welt.de/finanzen/immobilien/article111266698/Pfusch-am-Bau-ist-in-China-an-der-Tagesordnung.html.

Die Zeit, 2014. Deutschlandkarte - Trost für Berlin. *No. 12, 2014*. Available at: http://www.zeit.de/2014/12/deutschlandkarte-start-ups-berlin.

Die ZEIT, 2013. Führungskräfte: Selbstverliebte Chefs sind innovativer | Karriere | ZEIT ONLINE. Available at: http://www.zeit.de/karriere/beruf/2013-04/studie-selbst verliebte-chefs [Accessed April 8, 2013].

Die ZEIT, 2014. Wer gibt ihnen eine Chance? *10.02.2014*. Available at: http://www.zeit.de/2014/06/wagniskapital-start-up-gruender.

Digital Trends, 2012. Apple introduces "Blue Sky" program to give select employees time for personal projects. Available at: http://www.digitaltrends.com/apple/apple-introduces-blue-sky-program-to-give-select-employees-time-for-personal-projects/#!DrWjm.

Dorf, R.C., 1998. *The Technology Management Handbook*, CRC Press.

Dow Jones VentureSource, 2010. Dow Jones VentureSource. Available at: http://fis.dowjones.com/VS/4QUSFinancing.html.

Dowling, G., 2002. Customer Relationship Management: In B2C markets, often less is more. *CALIFORNIA MANAGEMENT REVIEW*, 44(3).

Dubini, P., 1989. Which venture capital backed entrepreneurs have the best chances of succeeding? *Journal of Business Venturing*, (4(2)), pp.123–132.

Dunbar, R., 1993. CO-EVOLUTION OF NEOCORTEX SIZE, GROUP SIZE AND LANGUAGE IN HUMANS. *Behavioral and Brain Sciences*, (16), pp.681–735.

Dweck, C., 2008. Can Personality Be Changed? The Role of Beliefs in Personality and Change - cdweckpersonalitychanged.pdf. *Association for Psychological Science*, 17(6), pp.391–394.

Ebben, J. & Johnson, A., 2006. Bootstrapping in small firms: An empirical analysis of change over time. *Journal of Business Venturing*, (21(6)), pp.851–865.

Ehrlich, S.B. & De Noble Tracy, A.F., 1994. After the cash arrives: a comparative study of venture capital and private investor involvement in entrepreneurial firms. *Journal of Business Venturing*, (9(1)), pp.67–82.

Eisenhardt, K. & Schoonhoven, C., 1990. Organizational growth: Linking founding team, strategy, environment, and growth among U.S. semiconductor ventures. *Administrative Science Quarterly*, 35, pp.504–529.

Elston, J.A. & Audretsch, D.B., 2010. Risk attitudes, wealth and sources of entrepreneurial start-up capital. *Journal of Economic Behavior & Organization*, (76(1)), pp.67–82.

Encyclopædia Britannica, 2014. Feudalism. Available at: http://www.britannica.com/EBchecked/topic/205583/feudalism.

Encyclopædia Britannica, Wirtschaftswunder. Available at: http://www.britannica.com/EBchecked/topic/645835/Wirtschaftswunder.

Entrepreneur.com, 2009. A Definitive Guide to Government Grants. Available at: http://www.entrepreneur.com/article/202110.

Entrepreneur.com, How to write a Business Plan. Available at: http://www.entrepreneur.com/businessplan/index.html [Accessed March 20, 2012].

Eulenberger, G., 2008. *Neue Finanzierungsformen in der Strategic Corporate Finance: Entrepreneurial Finance, Venture Capital, Private Equity und Hedge-Fonds.*, Springer Berlin Heidelberg.

European Commission, The EU Framework Programme for Research and Innovation. *Horizon 2020 - European Commission.* Available at: http://ec.europa.eu/programmes/horizon2020/.

European Commission DG Enterprise & Industry & ECORYS, 2011. Study on precommercial procurement in the field of Security Within the Framework Contract of Security. Available at: http://ec.europa.eu/enterprise/policies/security/files/doc/pcp_sec_finalreport_en.pdf [Accessed November 16, 2014].

European Institute of Innovation & Technology, 2014. EIT. Available at: http://eit.europa.eu/eit-community/eit-glance/eit-strategy-2014-2020.

European Investment Fund, 2011. *Business Angels in Germany,*

European Union, 2013. Erasmus for Entrepreneurs. Available at: http://www.erasmus-entrepreneurs.eu.

Europolitan, 2006. Deutschland und USA: Insolvenz-Verwaltung im Vergleich: Gestrauchelten eine Chance,. *Europolitan.* Available at: http://www.europolitan.de/cms/?s=ep_artikel&artikelid=1656&.

EVCA, 2010. Toolbox, Glossary. Available at: ttp://www.evca.eu/toolbox/glossary.aspx.

Exist, 2013. Exist Gründerstipendium. Available at: http://www.exist.de/exist-gruenderstipendium/index.php.

Fairlie, R.W., 2011. Entrepreneurship, Economic Conditions, and the Great Recession. Available at: http://economics.ucsc.edu/research/downloads/recessionentrep-v14.pdf.

Faitin, G., 2008. *Kopf schlägt Kapital,* Hanser.

FastCompany, 2013. Google Took Its 20% Back, But Other Companies Are Making Employee Side Projects Work For Them. Available at: http://www.fastcompany.com/3015963/google-took-its-20-back-but-other-companies-are-making-employee-side-projects-work-for-them.

FAZ, 2013. Die größten Unternehmen. *03.07.2013.* Available at: http://www.faz.net/aktuell/wirtschaft/unternehmen/rangliste-die-groessten-unternehmen-2013-12267817.html.

FAZ, 2014. Startups Search FAZ. *faz.net*. Available at: http://www.faz.net/suche/?query =startups&suchbegriffImage.x=0&suchbegriffImage.y=0&resultsPerPage=20.

FAZ.net - Börsenlexikon, 2014. Spin-off. *Börsenlexikon*. Available at: http://boersen lexikon.faz.net/spinoff.htm.

Federal Ministry of Family Affaires, 2013. Gute Kinderbetreuung. Available at: http://www.bmfsfj.de/BMFSFJ/Kinder-und-Jugend/kinderbetreuung.html.

Fisher III, W. & Oberholzer-Gee, F., 2013. Strategic Management of Intellect ual Property – An Integrated Approach. *California Management Review*. Available at: http://www.law.harvard.edu/faculty/faculty-workshops/fisher.faculty.workshop.summer-2013.pdf [Accessed October 25, 2014].

Fisher, P., *Building a Business: Raising Capital, Doing Deals*,

Florida, R., 2013. America's Leading Metros for Venture Capital. *CityLab.com*. Available at: http://www.citylab.com/work/2013/06/americas-top-metros-venture-capital/ 3284/.

Focus, 2013a. 52 deutsche Städte im Kita-Check. , (23). Available at: http://www.focus.de/finanzen/news/tid-31956/politik-endspurt-fuer-die-kindergar ten-garantie-52-deutsche-staedte-im-kita-check_aid_1021235.html.

Focus, 2013b. Deutsche Bürger sind bald Schulden-Europameister. , (29.08.2013). Available at: http://www.focus.de/finanzen/news/schlimmer-als-spanien-und-italien-deutsche-auf-dem-weg-zum-schulden-europameister_aid_1085143.html.

Foerderland, 2014. Wettbewerbe für Gründer Unternehmer und Management. Available at: http://www.foerderland.de/news/wettbewerbe/alle/awards/.

Forbes, 2013a. Airbnb And The Unstoppable Rise Of The Share Economy. Available at: http://www.forbes.com/sites/tomiogeron/2013/01/23/airbnb-and-the-unstoppable-rise-of-the-share-economy/2/.

Forbes, 2014a. Immigrant Entrepreneurs: Vital for American Innovation. *Techonomy*. Available at: http://www.forbes.com/sites/techonomy/2014/01/23/immigrant-entrepreneurs-vital-for-american-innovation/.

Forbes, 2014b. The World's Top 10 Venture Investors For 2014. Available at: http://www.forbes.com/sites/alexkonrad/2014/03/26/midas-top-ten-list-for-2014/.

Forbes, 2013b. U.S. Leads World In Burgeoning Crowdfunding Trend. Available at: http://www.forbes.com/sites/groupthink/2013/04/12/u-s-leads-world-in-burgeoning-crowdfunding-trend/.

Forbes, 2012. What's the Difference Between Private Equity and Venture Capital? *Forbes*. Available at: http://www.forbes.com/sites/victorhwang/2012/10/01/presidential-debate-primer-whats-the-difference-between-private-equity-and-ventu re-capital/.

Förderland, 2008. Business Angels: USA weit vor Deutschland. Available at: http://www.foerderland.de/419+M590c4e577e9.0.html [Accessed March 16, 2012].

Fortune, 2014. Fortune 500. *fortune.com*. Available at: http://fortune.com/fortune500/.

Frank, B., 2006. Anzeichen für nachhaltigen Aufschwung. *Venture Capital Magazin*, pp.38–39.

Frank, M.Z. & Goyal, V.K., 2007. *Trade-Off and Pecking Order Theories of Debt*, SSRN eLibrary.

Frazier, D., Heesen, M.G. & Taylor, J.S., 2011. National Venture Capital Association Yearbook 2011.

Frederick, H.H. & Kuratko, D.F., 2010. *Entrepreneurship : theory, process, practice*,

Freel, M.S., 2006. Are Small Innovators Credit Rationed? *Small Business Economics*, (28(1)), pp.23–35.

Freeman, J., 1996. Venture capital as an economy of time.

Friedmann, M., 1962. *Capitalism and Freedom*, University of Chicago Press.

Fried, V.H. & Hisrich, R.D., 1994. Toward a Model of Venture Capital Investment Decision Making. *Financial Management*, (23(3)), pp.28–37.

Gabler, 2014a. Definition » Bottom-up-Planung « | Gabler Wirtschaftslexikon. Available at: http://wirtschaftslexikon.gabler.de/Definition/bottom-up-planung.html [Accessed December 8, 2014].

Gabler, 2014b. Definition » Top-Down-Prinzip « | Gabler Wirtschaftslexikon. Available at: http://wirtschaftslexikon.gabler.de/Definition/top-down-prinzip.html [Accessed December 8, 2014].

Gabler, Due Dilligence. *Gabler Wirtschaftslexikon*. Available at: http://wirtschafts lexikon.gabler.de/Archiv/9219/due-diligence-v12.html.

Gabler, Keiretsu. *Gabler Wirtschaftslexikon*. Available at: http://wirtschafts lexikon.gabler.de/Definition/keiretsu.html [Accessed March 3, 2012b].

Gambles, I., 2009. *Making the Business Case - Proposals that Succeed for Projects that Work*, Gower Publishing Ltd.

Gartner, W.B., 1988. Who is an entrepreneur? Is the wrong question. *American journal of small business*, 12(4), pp.11–32.

Gartner, W.B., Starr, J.A. & Bhat, S., 1999. Predicting new venture survival:: An analysis of. *Journal of Business Venturing*, (14(2)), pp.215–232.

Gassmann, O. & Enkel, E., 2006. Open Innovation - Die Öffnung des Innovationsprozesses erhöht das Innovationspotenzial. *zfo Wissen*, (03/2006), pp.132–138.

Gaul, W., 2002. *Marketing & Innovation*,

Gelfert, H.-D., 2011. http://www.tagesspiegel.de/meinung/die-widerspruechlichen-usa/5803388.html. *Der Tagesspiegel*. Available at: http://www.tagesspiegel.de/meinung/die-widerspruechlichen-usa/5803388.html.

George, G. & Bock, A., 2011. The Business Model in Practice and its Implications for Entrepreneurship Research. *Entrepreneurship Theory and Practice*, 35(1), pp.83–111.

German Trade & Invest (GTAI), 2013. Coming up Trumps. *German Trade & Invest*. Available at: http://www.gtai.de/GTAI/Navigation/EN/Invest/Service/Publications/markets-germany,did=600136.html.

German/ US Venture Capitalists, 2010. Venture Capital in Germany & the US.

Gifford, S., 1997. Limited attention and the role of the venture capitalist. *Journal of Business Venturing*, (12(6)), pp.459–482.

Global Entrepreneurship Monitor, 2011. *Global Entrepreneurship Monitor, Unternehmensgründungen im weltweiten Vergleich*,

Global Entrepreneurship Monitor, 2012. Global Entrepreneurship Monitor - Unternehmensgründungen im weltweiten Vergleich.

Global Entrepreneurship Monitor, 2013. *Global Entrepreneurship Monitor, Unternehmensgründungen im weltweiten Vergleich*,

Gneezy, U., 2002. Gender & Competition. Available at: http://www.chicagobooth.edu/capideas/fall02/genderandcompetition.html.

Golley, F., 1993. A History of the Ecosystem Concept in Ecology. More than the Sum of the Parts. *Yale University Press*.

Gompers, P.A. & Lerner, J., 2004. *The Venture Capital Cycle*, MIT Press.

Gorman, M. & Sahlmann, W.A., 2010. What do venture capitalists do? *Journal of Business Venturing*, (30(3)), pp.207–214.

Gov UK, 2012. Bankruptcy: A Fresh Start. *The Insolvency Service*. Available at: http://www.insolvencydirect.bis.gov.uk/insolvencyprofessionandlegislation/con_do c_register/con_doc_archive/consultation/freshstart/sec4.htm.

Gründerszene, 2013. Unternehmertum. Available at: http://www.gruenderszene.de/lexikon/begriffe/unternehmertum.

GTAI, 2014. Wirtschaftsstandort Deutschland. Available at: http://www.gtai.de/GTAI/Content/DE/Invest/_SharedDocs/Downloads/GTAI/Sonstiges/blg-kurzueberblick-wirtschaftsstandort-deutschland-pdf.pdf [Accessed January 8, 2015].

Gurdon, M.A. & Samsom, K.J., 2010. A longitudinal study of success and failure among scientist-started ventures. *Technovation*, (30(3)), pp.207–214.

Haas, H., 2002. *Körperschaftsteuer* 7. ed., C.H. Beck.

Haas, I., 2008. *Die neue Abgeltungssteuer* 1st ed., Haufe.

Habeeb, A Brief About Internet Affiliate Marketing. *Consumer Electronic Magazine*. Available at: http://consumerelectronicsmagazine.com/a-brief-about-internet-affiliate-marketing/ [Accessed October 3, 2012].

Hage, S., 2007. Dann kopieren wir mit Stolz. *Manager Magazin*. Available at: http://www.manager-magazin.de/unternehmen/it/a-506291.html.

Hall, J. & Hofer, C.W., 1993. Venture capitalists' decision criteria in new venture evaluation. *Journal of Business Venturing*, (8(1)), pp.25–42.

Hambrick, D. & Meinz, E., 2011. Sorry, Strivers: Talent Matters. *New York Times*. Available at: Sorry, Strivers: Talent Matters.

Hamel, G., 2007. The Future of Management. *Harvard Business School Press*.

Handelsblatt, 2014a. Amazon dominiert deutschen Online-Handel. *Handelsblatt*. Available at: http://www.handelsblatt.com/unternehmen/handel-dienstleister/riesen-vorsprung-amazon-dominiert-deutschen-online-handel/9948286.html.

Handelsblatt, 2014b. Gründer Special - Mit kleinen Geschäften groß rauskommen. *Handelsblatt.de*. Available at: http://www.handelsblatt.com/unternehmen/mittelstand/special-existenzgruendung/start-up-des-monats-sugartrends-mit-klein en-geschaeften-gross-rauskommen/10846514.html.

Hanley, A. & Girma, S., 2006. New Ventures and their Credit Terms. *Small Business Economics*, (26(4)), pp.351–364.

Häring, N. & Storbeck, O., 2007. *Ökonomie 2.0*, Schäffer, Poeschel.

Haussschildt, J., 1999. *Promotoren – Champions der Innovation*, Gabler.

Haussschildt, J. & Salomo, S., 2007. *Innovationsmanagement*, Vahlen.

Heckmann, J.J., 2009. Invest in early childhood development: : Reduce deficits, strengthen the economy. Available at: http://www.actforchildren.org/ site/DocServer/Heckman_Fact_Sheet.pdf?docID=1169.

Heisenberg, W., 1927. Über den anschaulichen Inhalt der quantentheoretischen Kinematik und Mechanik. , pp.172–198.

Henn, C., 2008. Lizenz zum Blutsaugen. *Stern*, (21. Januar 2008). Available at: http://www.stern.de/digital/online/abmahnungen-lizenz-zum-blutsaugen-607480.html.

Hersch, J., 1989. *Das philosopische Staunen*, Piper.

Himmelrath, A., 2012. Wo ist hier das Druckventil? *KarriereSpiegel*. Available at: http://www.spiegel.de/karriere/berufsstart/bachelor-studenten-wo-ist-hier-das-druckventil-a-811920.html.

Hipple, S., 2010. Self-employment in the United States. *bis.gov*. Available at: http://www.bls.gov/opub/mlr/2010/09/art2full.pdf.

HM Revenue & Customs, 2013. *Enterprise Investment Scheme*, Available at: http://www.hmrc.gov.uk/eis/part1/1-2.htm#121.

Hoffmann, A., 2013. Heilemann-Brüder nehmen DailyDeal zurück. *Gründerszene*. Available at: http://www.gruenderszene.de/news/heilemann-dailydeal-google-zuruck.

Hofmann, A., 2014. „Wir brauchen eine Börse, die Wachstums-Unternehmen für IPOs gewinnen kann". *Gründerszene*. Available at: http://www.gruenderszene.de/ allgemein/gabriel-boerse.

Hoggson, N., 2007. *Banking Through the Ages: From the Romans to the Medicis, from the Dutch to the Rothschilds*, Cosimo, Inc.

Hoopes, J., 2003. *False Prophets - The Gurus Who Created Modern Management and Why Their Ideas Are Bad for Business Today*, Basic Books.

Horváth, Ö. v., 1937. *Der jüngste Tag*, Reclam.

Hubbard, G. & Dudley, W., 2004. *How Capital Markets Enhance Economic Performance and Facilitate Job Creation*, Columbia University & Global Markets Institutes, Goldman Sachs. Available at: http://www0.gsb.columbia.edu/faculty/ghubbard/ Articles%20for%20Web%20Site/How%20Capital%20Markets%20Enhance%20Economic%20Performance%20and%20Facilit.pdf.

Huffington Post, 2014. "Deutschland, armes Gründerland" - Das giftige Klima für deutsche Startups. *The Huffington Post.* Available at: http://www.huffingtonpost.de/2014/08/28/deutschland-armes-gruenderland_n_5727588.html.

Ifex, 2010. *Ifex Gründungen,* Stuttgart.

Ifex, 2014. Innovationgutschein BW. Available at: https://www.innovationsgut scheine.de/de/Foerderbedingungen.php.

Inc., 1996. Fleshes of Genius. Available at: http://www.inc.com/magazine/19960515/2083_pagen_5.html.

Inglehart, R. & Welzel, C., 2005. *Modernization, Cultural Change, and Democracy: The Human Development Sequence,* Cambridge University Press.

Innovationsallianz NRW, 2014. Innovationsgutschein NRW.

Insolvenzrecht.info, 2012. Die Restschuldbefreiung. *Insolvenzrecht.info.* Available at: http://www.insolvenzrecht.info/verbraucherinsolvenz/restschuldbefreiung.html.

Institut für Mittelstandsforschung, 2012. Gründungen, Liquidationen, Insolvenzen 2010 in Deutschland. Available at: http://www.ifm-bonn.org//uploads/tx_ifmstudies/Daten-und-Fakten-1_2011.pdf.

International Monetary Fund (IMF) & Mihet, Roxana, 2012. Effects of Culture on Firm Risk-Taking: A Cross-Country and Cross-Industry Analysis. Available at: https://www.imf.org/external/pubs/ft/wp/2012/wp12210.pdf.

Internetworld, 2013. Break-even noch in diesem Jahr? *Internetworld.de.* Available at: http://heftarchiv.internetworld.de/2013/Ausgabe-22-2013/Break-even-noch-in-diesem-Jahr.

Investopedia, 2009. How did dotcom companies become so overvalued in the late 1990s? *Investopedia.* Available at: http://www.investopedia.com/ask/answers/08/dotcom-overvalued-valinux.asp.

Investopedia, 2014. Investopedia - Series A. Available at: http://www.investopedia.com/terms/s/seriesa.asp.

IRS, 2013. IRS - Capital Gains & Losses. Available at: http://www.irs.gov/taxtopics/tc409.html.

Isenberg, D., 2011. Introducing the Entrepreneurship Ecosystem: Four Defining Characteristics. *Forbes Magazin.*

Iwata, E., 2012. Entrepreneur Guy Kawasaki doesn't accept failure. *USA Today.* Available at: http://abcnews.go.com/Business/story?id=6218413.

Jacobi, O., 2013. Die deutsche Venture-Capital-Szene – Zahlen, bitte! *Gründerszene*. Available at: http://www.gruenderszene.de/finanzen/deutsche-venture-capital-szene-statistiken.

Jacobi, O., 2010. Kolumne: Die deutsche Venture-Capital-Szene – Wie alles begann -. Available at: http://www.gruenderszene.de/finanzen/kolumne-die-deutsche-venture-capital-szene-%e2%80%93-wie-alles-begann [Accessed March 4, 2012].

Jacobsen, K.L., 2003. *Bestimmungsfaktoren für Erfolg im Entrepreneurship*. Freie Universität Berlin.

Janssen, F., 2006. Do Managers' Charateristics influence Employment Growth of SMEs? *Journal of Small Business and Entrepreneurship*, pp.293–315.

Jax, K., 2004. Haben Ökosysteme eine Eigenart? Gedanken zur Rolle des Eigenart-Begriffs in naturwissenschaftlich geprägten Naturschutzdiskussionen - HamburgUP_Projektionsflaeche_Jax.pdf. Available at: http://hup.sub.uni-hamburg.de/volltexte/2008/74/chapter/HamburgUP_Projektionsflaeche_Jax.pdf [Accessed October 10, 2014].

Jones, G., 2013. *Entrepreneurship and Multinationals: Global Business and the Making of the Modern World*,

Jones, M., 2005. *Superlatives USA: The Largest, Smallest, Longest, Shortest, and Wackiest Sites in America*, Capital Books Inc.

Jong, J. de & Wennekers, S., 2008. Intrapreneurship: Conceptualizing entrepreneurial employee behaviour.

Junge Innovatoren, 2013. Junge Innovatoren. Available at: http://www.junge-innovatoren.de/foerderung.html.

Kacperczyk, A.J., 2012. Social Influence and Entrepreneurship: The Effect of University Peers on Entrepreneurial Entry.

Kaczmarek, J., 2011. CityDeal – Der Milliarden-Exit ist anvisiert. *Gründerszene*. Available at: http://www.gruenderszene.de/exits/citydeal-groupon.

Kakati, M., 2003. Success criteria in high-tech new ventures. *Technovation*, (23(5)), pp.447–457.

Karl-Steinbuch-Stipendium, 2013. Steinbuch Stipendium. Available at: http://www.karl-steinbuch-stipendium.de/kss_stipendium.html.

Kauffman Foundation, 2011. OVERCOMING THE GENDER GAP: WOMEN ENTREPRENEURS AS ECONOMIC DRIVERS.

Keightley, M. & Sherlock, M., 2014. The Corporate Income Tax System: Overview and Options for Reform. Available at: https://www.hsdl.org/?view&did=760400.

Kelly, N., 2014. Will Spanish Help You Reach the U.S. Hispanic Market? It Depends. *Harvard Business Review*. Available at: http://blogs.hbr.org/2014/02/will-spanish-help-you-reach-the-u-s-hispanic-market-it-depends/.

Kepler, E. & Shane, S., 2007. Are Male and Female Entrepreneurs Really That Different?

KfW, 2013a. ERP Startfonds. Available at: https://www.kfw.de/inlandsfoerderung/Unternehmen/Gr%C3%BCnden-Erweitern/Finanzierungsangebote/ERP-Startfonds-%28136%29/.

KfW, 2013b. KfW Gründercoaching. Available at: https://www.kfw.de/inlandsfoer derung/Unternehmen/Unternehmen-erweitern-festigen/Finanzierungsangebote/Gr %C3%BCndercoaching-Deutschland-%28GCD%29/.

KfW-Research, 2012. KfW Gründungsmonitor 2012.

KIT, 2014. Netzwerk am KIT. Available at: http://kit-gruenderschmiede.de/de/netzwerk/am-kit/.

Kocieniewsk, D., 2012. Zuckerberg's Big Tax Bill May Benefit Facebook. *New York Times*. Available at: http://www.nytimes.com/2012/02/04/business/zuckerbergs-big-tax-bill-may-benefit-facebook.html.

Kohn, J., 2013. Verletze US Veteranen. Available at: http://www.spiegel.de/politik/ausland/verletztendrama-um-us-kriegsrueckkehrer-der-wundenheiler-a-877689.html [Accessed January 28, 2013].

Kollmann, T.P.D., 2006. *E-Entrepreneurship*,

Komisar, R, 2010. Are Business Plans a Work of Fiction? Available at: http://ecorner.stanford.edu/authorMaterialInfo.html?mid=2416 [Accessed March 28, 2012].

Kontratjew, N., 1926. *Die langen Wellen der Konjunktur*, Heptagon.

Krauße, A.-C, 2005. *Geschichte der Malerei*, Tandem.

Kreft, V., 2011a. *Einkommenssteuerrecht*, Alpmann Schmidt.

Kreft, V., 2011b. *Einkommenssteuerrecht* 12th ed., Alpmann Schmidt. Available at: http://www.buecher.de/shop/einkommensteuer-lohnsteuer-und-kapitalertragsteue r-kirchensteuer/einkommensteuerrecht/kreft-volker/products_products/detail/pr od_id/32539858/ [Accessed February 14, 2012].

Kulke, U., 2011. Diese deutsche Kleinstaaterei war segensreich. *Die Welt*. Available at: http://www.welt.de/kultur/history/article13641035/Diese-deutsche-Kleinstaaterei-war-segensreich.html.

Kuratko, D.F., 2005. *The Emergence of Entrepreneurship Education: Development, Trends, and Challenges*, Baylor University.

Landström, H., 2007. *Handbook of research on venture capital,*

Larsson, J. & Rooswall, M., 2001. *Overruling uncertainty-A study of venture of capital decision making,*

Lee, J., Gino, F. & Staats, B., 2012. Rainmakers: Why Bad Weather Means Good Productivity. Available at: http://dash.harvard.edu/bitstream/handle/1/9299650/13-005.pdf?sequence=1.

Leibenstein, H., 1968. The theory of economic development: an inquiry into profits, capital, credit, interest, and the business cycle. *The American Economic Review*, 58(2), pp.72–83.

Lester, R. & Piore, M., 2004. *Innovation - The missing dimension*, Harvard University Press.

Lister, K., 2011. The Myth of the Business Plan. Available at: http://www.entrepreneur.com/article/220440 [Accessed March 27, 2012].

Lloyd, S., 2013. David Rock Drops Some Knowledge About 360 Assessments and SCARF 360 for Developing Leaders. *Qualtrics Blog*. Available at: http://www.qualtrics.com/blog/360-assessments/.

Loayza, J., 2013. The pros and cons of going into business with a friend. *The Global and Mail*. Available at: http://www.theglobeandmail.com/report-on-business/small-business/starting-out/the-pros-and-cons-of-going-into-business-with-a-friend/article14086769/.

Locke, J., 1690. *An Essay Concerning Human Understanding*, Scolar Press.

Luukkonen, T. & Maunula, M., 2006. Coaching' Small Biotech Companies into Success: The Value-adding Function of VC.

Luukkonen, T. & Maunula, M., 2007. Non-financial Value-added of Venture Capital: A Comparative Study of Different Venture Capital Investors.

Mace, G., 1979. *Locke, Hobbes, and the Federalist Papers: An Essay on the Genesis of the American Political Heritage*, Southern Illinois University Press.

Machiavelli, N., 1513. *Il Principe,*

Macht, S., 2006. The post-investment impact of Business Angels upon their investee companies. In The post-investment impact of Business Angels upon their investee companies. Cardiff.

MacMillan, I.C., Siegel, R. & Subba Narasimha, P.N., 1985. Criteria used by venture capitalists to evaluate new venture proposals. *Journal of Business Venturing*, (1(1)), pp.119–128.

MacMillan, I.C., Zemann, L. & Subba Narasimha, P.N., 1987. Criteria distinguishing successful from unsuccessful ventures in the venture screening process. *Journal of Business Venturing*, (2(2)), pp.123–137.

Mader, S., 2011. *Concepts of Biology* 2nd ed., McGraw Hill.

Magnusson, L., 2002. *Mercantilism: The Shaping of an Economic Language*, Taylor & Francis.

Maier, M., 2011a. Start-ups in Deutschland und den USA. *Analysen & Argrumente*, (99), p.11.

Maier, M., 2011b. Start-ups in Deutschland und den USA. Analysen & Argrumente. *Konrad Adenauer Stiftung*, (99).

Majunke Consulting, 2013. VENTURE CAPITAL - Yearbook 2012.

Manger, R., 2011. Investing in startups.

Marquardt, M.J., 2005. *Globalization: The Pathway to Prosperity, Freedom and Peace*, George Washington University.

Marshall, M., 2008. The German startup scene - copycats, but getting smarter. Available at: http //venturebeat.com/2008/10/22/the-german-start-up-scene-copycats-but-getting-smarter/.

Martens, R., 1996. *Successful Coaching* 4th ed.,

Mathur, A., 2011. *Beyond Bankruptcy: Does the Bankruptcy Code Provide a Fresh Start To Entrepreneurs?*, Small Business Administration. Available at: https://www.sba.gov/sites/default/files/Bankruptsy%20Report.pdf.

Mayer, S.E., 2002. The Influence of Parental Income on Children's Outcomes. Available at: http://www.msd.govt.nz/documents/about-msd-and-our-work/publications-resources/research/influence-parental-income/influence-of-parental-income.pdf [Accessed January 5, 2015].

McDougal, P.P., 1989. *International vs. domestic entrepreneurship*,

McDougal, P.P., Oviatt, B.M. & Shrades, R.C., 2003. *A comparison of international and domestic new ventures, Journal of international entrepreneurship,*

McFarlane, D., 2013. The Importance of Business Ethics to Small Ventures. *Entrepreneurship and Innovation Management Journal,* 1(1), pp.50–59.

Melendez, R., 2008. Five Elements to Include in a Compelling Business Case. *Industry Week.* Available at: http://www.industryweek.com/articles/five_elements_to_include _in_a_compelling_business_case_15594.aspx.

Merriam-Webster, 2010a. Entrepreneur -Definition. *Free Merriam-Webster Dictionary.* Available at: http://www.merriam-webster.com/dictionary/entrepreneur [Accessed December 28, 2010].

Merriam-Webster, 2010b. Innovation - Definition. *Free Merriam-Webster Dictionary.* Available at: http://www.merriam-webster.com/dictionary/innovation [Accessed December 28, 2010].

Merriam-Webster, 2010c. Success - Definition and More. *Free Merriam-Webster Dictionary.* Available at: http://www.merriam-webster.com/dictionary/entrepreneur [Accessed September 11, 2010].

Merriam-Webster, 2011. System. *Merriam-Webster.* Available at: http://www.merriam-webster.com/dictionary/system.

Meza, D. & Southey, C., 1996. The Borrower's Curse: Optimism, Finance and Entrepreneurship. *The Economic Journal,* (106(435)), pp.375–386.

Miller, W.I., 1993. *Humiliation: and other essays on honor, social discomfort, and violence,* Cornell University Press.

Ministry of Economy State of Israel, 2014. Homeland Security (HLS) & Public Safety in Israel. *Invest in Israel.* Available at: http://www.investinisrael.gov.il/ NR/exeres/7C2F6937-A259-4A4A-9C29-DE351032B87A.htm.

MIT, 2014. Martin Trust - Center for Entrepreneurship. Available at: https://entre preneurship.mit.edu/community/overview/.

MIT $100k, 2011. MIT $100K Competition. Available at: http://www.mit100k.org/.

Moore, J.F., 1993. Predators and prey: a new ecology of competition. *Harvard Business Review,* (71(3)), pp.75–86.

Morris, G., 2012. Average Venture Capital Deal Sizes Globally in 2012 – December 2012. *Preqin.* Available at: https://www.preqin.com/blog/101/6044/vc-deals-2012.

Mulcahy, D., 2005. *Venturing forward - A practical guide to raising equity capital in Ireland*, Oak Tree Press.

Müller-Prothmann, T. & Dörr, N., 2014. *Innovationsmanagement: Strategien, Methoden und Werkzeuge für systematische Innovationsporzesse*, Hanser.

Müller, R., 1997. *Innovation gewinnt*, Zürich: Orell Füssli.

Myers, S.C. & Majluf, N.S., 1984. Corporate financing and investment decisions when firms have informationthat investors do not have. *NBER Working paper*.

Nagel, C., 2011. Interview mit einem VC: Earlybird. Available at: http://www.gruen derszene.de/finanzen/christian-nagel-earlybird-vc-venture-capital.

National Philantropic Trust, 2013. *Charitable Giving Statistics*, Available at: http://www.nptrust.org/philanthropic-resources/charitable-giving-statistics/.

Nationmaster, 2014. English speaking countries: Statistical Profile. *Nationmaster.com*. Available at: http://www.nationmaster.com/country-info/groups/English-speaking-countries.

Nationsonline.org, 2014. Most widely spoken Languages in the World. *Nationsonline.org*. Available at: http://www.nationsonline.org/oneworld/most_spoken_languages.htm.

Neck, H. et al., 2004. An Entrepreneurial System View of New Venture Creation. *Journal of Small Business Management*, (42(2)), pp.190–208.

Neeley, T., 2012. Global Business Speaks English. *Harvard Business Review*, (May). Available at: https://hbr.org/2012/05/global-business-speaks-english/ar/1.

Newton, D., 2001. Understanding the Financing Stages. Available at: http://www.entrepreneur.com/article/42336.

New York Times, 2014. Startups Search NYTimes. *njtimes.com*.

Noelle-Neumann, E., 1996. *Öffentliche Meinung*, Ullstein.

OECD, 2001. *Fostering high-tech spin-offs: a public strategy for innovation*, Paris: OECD Publishing.

OECD, 2003. *The Sources of Economic Growth*, OECD Publishing.

OECD & Eurostat, 2005. *The Oslo Manual - GUIDELINES FOR COLLECTING AND INTERPRETING INNOVATION DATA*,

Olufsson, G.L., 2001. Early stage financing of NTBFs: an analysis of contributions from support actors. *Venture Capital: An International Journal of Entrepreneurial Finance*, (3(2)), pp.151–168.

Organisation of Economic Cooperation and Development (OECD), 2011. *Financing High-Growth Firms | The Role of Angel Investors*, OECD Publishing. Available at: http://dx.doi.org/10.1787/9789264118782-en.

Orthiese, J., 2007. *Value added by venture capital firms : an analysis on the basis of new technology-based firms in the USA and Germany*, Köln: Lohmar.

Ortmans, J., 2009. Infrastructure & Entrepreneurship. *entrepreneurship.org*. Available at: http://www.entrepreneurship.org/policy-forum/infrastructure-and-entrepreneurship.aspx.

Osnabruegge, M. van, 2000. A comparison of business angel and venture capitalist investment procedures: an agency theory-based analysis. *Venture Capital: An International Journal of Entrepreneurial Finance,*, (2(2)), pp.91–109.

Osterwalder, A., Pigneur, Y. & Smith, A., 2010. *Business Model Generation; A handbook for visionaries, game changers and challengers*, John Wiley & Sons.

Oxford Dictionaries, 2014. Experience. *Oxford Dictionaries.* Available at: http://www.oxforddictionaries.com/definition/english/experience.

Paris, B.J.M., Arndt, M. & Berner, R., 2006. The World's Most Innovative Companies.

Paymetric, 2013. Paymetric Selected as a 2013 Red Herring Top 100 Global Compan. Available at: http://www.paymetric.com/news/paymetric-selected-2013-red-herring-top-100-global-company.

Payne, G.T., 2009. The Deal Structuring Stage of the Venture Capitalist Decision-Making Process: Exploring Confidence and Control. *Journal of Small Business Management*, (47(2)), pp.154–179.

Pepels, W., 2004. *Handbuch des Marketing* 4th ed., Oldenbourg Wissenschaftsverlag.

Perkins, T., 2007. Global Business.

Perner, L., 1999. Group Influence. Available at: http://www.consumerpsychologist.com/cb_Group_Influences.html.

Pierenkemper, T. & Tilly, R., 2004. *The German Economy during the Nineteenth Century*, Berghahn Books.

Pinchot III, G. & Pinchot, E.S., 1978. Intra-Corporate Entrepreneurship. Available at: http://www.intrapreneur.com/MainPages/History/IntraCorp.html [Accessed March 28, 2012].

Piper, N., 2012. Steuerforderung an Facebook-Gründer. *Süddeutsche Zeitung (SZ)*. Available at: http://www.sueddeutsche.de/wirtschaft/steuerforderung-an-facebook-gruender-spitzensatz-fuer-zuckerberg-1.1276233.

Pirner, H., 2007. Mit der Unbestimmtheit rechnen.

Pliz, G., 2009. *Networking*, Deutsche Taschenbuch Verlag.

Poets & Quants, 2013. Top 100 MBA Startups. *Poets & Quandts*. Available at: http://poetsandquants.com/2013/11/18/poetsquants-top-100-mba-startups/2/.

Pozin, I., 2012. 10 Startups Changing the World... *Forbes*. Available at: http://www.forbes.com/sites/ilyapozin/2012/05/09/10-startups-changing-the-world-and-what-we-can-learn-from-them/.

Precht, R.D., 2013. *Anna die Schule und der liebe Gott*, Goldmann.

Preqin, 2013. *Venture Capital Deals Activity Continues Slide in Q4 2012*, Available at: https://www.preqin.com/docs/press/Venture_Capital_Q4_2012.pdf.

Preuß, R. & Osel, J., 2012. Harsche Kritik an Bachelor und Master. *Süddeutsche Zeitung (SZ)*. Available at: http://www.sueddeutsche.de/bildung/zehn-jahre-bologna-reform-harsche-kritik-an-bachelor-und-master-1.1441136.

Prof. Dr. Backes-Gellner, U. et al., 2012. *Gutachten zu Forschung, Innovation und technologischer Leistungsfähigkeit in Deutschland 2012*, Expertenkommission Forschung und Innovation (EFI). Available at: http://www.stifterverband.de/statistik_und_analysen/wissenschaftsstatistik/efi/efi_jahresgutachten_2012.pdf [Accessed March 3, 2012].

Public Broadcasting Service, Andrew Carnegie's life. Available at: http://www.pbs.org/wgbh/amex/carnegie/filmmore/description.html.

PWC, 2013. Vermittlungsausschuss beendet Ungleichbehandlung in der Besteuerung von Streubesitzdividenden. Available at: http://www.pwc.de/de/steuerberatung/vermittlungsausschuss-beendet-ungleichbehandlung-in-der-besteuerung-von-streu besitzdividenden.jhtml.

PWC & NVA, 2013. *ANNUAL VENTURE INVESTMENT DOLLARS DECLINE FOR FIRST TIME IN THREE YEARS, ACCORDING TO THE MONEYTREE REPORT*,

De Raad, B., 1998. Five big, big five issues: Rationale, content, structure, status and crosscultural assessment. , European Psychologist(3), pp.113–124.

Ranade, V., 2008. Early-Stage Valuation in the Biotechnology Industry.

Räth, G., 2013. HTGF II knackt 300-Millionen. *Gründerszene*. Available at: http://www.gruenderszene.de/news/htgf-metro-media-more-venture.

RedHerring, 2011. 2011 Top 100 Global Companies. *Redherring Top 100*. Available at: http://www.redherring.com/events/red-herring-global/rhg2013winners/.

RedHerring, 2012. 2012 Top 100 Global Companies. *Redherring Top 100*. Available at: http://www.redherring.com/events/red-herring-global/rhg2013winners/.

RedHerring, 2013. 2013 Top 100 Global Companies. *Redherring Top 100*. Available at: http://www.redherring.com/events/red-herring-global/rhg2013winners/.

Reich, R., 1987. Entrepreneurship reconsidered: The team as hero. *Harvard Business Review*, (May/June), pp.77–83.

Reuters, 2013a. Factbox: Apple, Amazon, Google and tax avoidance schemes. *Reuters*. Available at: http://www.reuters.com/article/2013/05/22/us-eu-tax-avoidance-idUSBRE94L0GW20130522.

Reuters, 2013b. Seqoia Capital raises 1.17 billion for new funds.

Reuters, 2011. The New Peer Pressure: At Stanford, "If You Haven't Started Company by Age 20, You're a Failure." *Pe Hub*.

Ripsas, S., 1997. *Entrepreneurship als ökonomischer Prozeß - Perspektiven zur Föderung unternehmerischen Handelns,*. Freie Universität Berlin.

Riquelme, H. & Rickards, T., 1992. Hybrid conjoint analysis: An estimation probe in new venture decisions. *Journal of Business Venturing*, (50(1)), pp.505–518.

Robinson, R.B., 1987. Emerging strategies in the venture capital industry. *Journal of Business Venturing*, (2(1)), pp.53–77.

Rock, D., 2008. SCARF: a brain-based model for collaborating with and influencing others. *NeuroLeadership journal*.

Roland Berger, 2014. Online automotive parts sales: The rise of a new channel. *www.rolandberger.com*. Available at: http://www.rolandberger.com/media/publications/2014-05-09-rbsc-pub-Online_automotive_parts_sales.html.

Romans, A., 2013. *The Entrepreneurial Bible to Venture Capital*, McGraw Hill.

Rothbard, M., 2006. *An Austrian Perspective on the History of Economic Thought Autor*, Ludwig von Mises Institute.

Rothschild, M., 1995. *Bionomics: Economy As Ecosystem*, Owl Books.

Roxbergh, C. & McKinsey & Company, 2009. The use and abuse of scenarios. Available at: http://www.mckinsey.com/insights/strategy/the_use_and_abuse_of_scenarios.

Rüdisüli, R., 2005. *Value creation of spin-offs and carve-outs*, Difo-Druck GmbH.

Ruhnka, J.F., Feldmann, H.D. & Dean, T.J., 1992. The "living dead" phenomenon in venture capital investments. *Journal of Business Venturing*, (7(2)).

RWTH Aachen, 2013. Avigle. Available at: http://www.fsd.rwth-aachen.de/English/ Research/Avigle.php [Accessed January 28, 2013].

Sächsische Aufbaubank, 2013. Seed Stipendium.

Sandberg, W.R. & Hofer, C.W., 1987. Improving new venture performance: The role of strategy, industry structure, and the entrepreneur. *Journal of Business Venturing*, (2(1)), pp.5–28.

Saridakis, G., Mole, K. & Hay, H., 2008. Do Liquidity Constraints in the First year of Trading Reduce the Likelihood of Firm Growth and Survival? Evidence from England.

Saublens, C. & Secretariat, E., 2008. Introduction to business angels and business angels network activities in Europe. Available at: www.fnaba.org.

Sawhney, R., 2011. Indian Venture Capital Scene.

Say, J.-B., 1821. *A Treatise on political economy*,

Schedlbauer, M., 2011. Making the Business Case. Available at: http://hartford. iiba.org/download/Making%20the%20Business%20Case.pdf.

Schumpeter, J.A., 1939. *Business Cycles, a theoretical, historical, and statistical analysis of the capitalist process*, Northwestern University.

Schumpeter, J.A., 1947. *Kapitalismus, Sozialismus, Demokratie* 8. Auflage 2005., A. Francke.

Schumpeter, J.A., 1911. *The theory of economic development: an inquiry into profits, capital, credit, interest, and the business cycle*, New Brunswick N.J.: Transaction Books.

Schurenberg, E., 2012. *Inc.* Available at: http://www.inc.com/eric-schurenberg/the-best-definition-of-entepreneurship.html.

Schwarzkopf, H.P.D., 2010. Hand of Fortune.

Schwarzkopf, H.P.D., 2009. International Business Law.

SEN, 2014. Stanford Entrepreneurship Network. Available at: http://sen.stanford.edu/.

Shaffer, D., 2009. *Social and personality development* 6th ed., Wadsworth.

Shepherd, D.A., 1999. Venture capitalists' assessment of new venture survival. *Management Science*, (45(5)), pp.621–632.

Silva, J., 2004. Venture capitalists' decision-making in small equity markets: a case study using participant observation. *Venture Capital: An International Journal of Entrepreneurial Finance*, (6(2-3)), pp.125–145.

Simmons, B.L., 2009. Entrepreneurs and the "Big five." Available at: http://www. bretlsimmons.com/2009-07/entrepreneurs-and-%E2%80%9Cthe-big-five%E2%80 %9D/.

Slavet, J., 2013. The surest way to build a billion-dollar company. *LinkedIn*. Available at: http://www.linkedin.com/today/post/article/20130403230119-58666-the-surest-way-to-build-a-billion-dollar-internet-company.

Smith, A., 1776. *The Wealth of Nations*,

Spiegel, 1995. Musik, Lust fürs Ohr. , (Spiegel Special 12/1995). Available at: http://www.spiegel.de/spiegel/spiegelspecial/d-9259878.html.

Spiegelonline, 2014a. Israelische Start-up-Gründer: Lieber Berlin als Tel Aviv. Available at: http://www.spiegel.de/wirtschaft/unternehmen/israelische-start-ups-in-berlin-a-953579.html.

Spiegelonline, 2014b. Karriere Spiegel. *spiegelonline*. Available at: http://www. spiegel.de/thema/gruenderzeit_karrierespiegel/.

Spiegelonline, 2013. Umstrittener Talkshow-Gast bei Jauch: Maschmeyer-Einladung sorgt für Ärger im NDR. Available at: http://www.spiegel.de/kultur/tv/guenther-jauch-maschmeyer-einladung-sorgt-fuer-aerger-im-ndr-a-887938.html.

Spilling, O., 1996. The Entrepreneurial System – On Entrepreneurship in the Context of a Mega Event. *Journal of Business Research*, (36(1)), pp.91–103.

Spinelli, S. & Adams, R., 2012. *New Venture Creation:Entrepreneurship for the 21st Century*, McGraw Hill.

Stanford University, 2008. Marketing and The Sales Pipeline, Entrepreneurship Course. Available at: http://web.stanford.edu/class/e145/2008_fall/protected/handouts/ E145_0807_steve_mktg.pdf.

Start2Grow, 2014. Start2Grow Business Plan. Available at: http://start2grow. de/de/gruendungswettbewerb_2015/businessplan.jsp.

Stevenson, H., 1983. *A perspective on entrepreneurship*, Harvard Business Review.

Stevenson, H., 2000. WHY ENTREPRENEURSHIP HAS WON!

Steward, W.H.J., 1998. A proclivity for entrepreneurship: A comparision of entrepreneur s, small business owners, and corporate managers. , Journal of Business Venturing 14.

Stowasser, J.M., Petschenig, M. & Skutsch, F., 1998. *Lateinisches-Deutsches Schulwärterbuch,*

Strauss, K., 2012. Airbnb's Explosive Growth Draws Big Cash (and a big valuation). *Forbes.* Available at: http://www.forbes.com/sites/karstenstrauss/2012/09/28/airbnbs-explosive-growth-draws-big-cash-and-a-big-valuation/.

Struck, U., 1990. *Geschäftspläne: Voraussetzung für erfolgreiche Kapitalbeschaffung,* Schäffer, Poeschel.

Stucky, W.P.D., 2013. Gründen aus der Universität heraus.

Study in the USA, 2012. Understanding the American Education System. *StudyUSA.com.* Available at: http://studyusa.com/en/a/58/understanding-the-american-education-system.

Süddeutsche Zeitung, 2011. Top 100 in Deutschland. *13.08.2011.*

Susanne, P., 2006. Vererbung und Erziehung: Wie Eltern ihre Kinder prägen - SPIEGEL ONLINE. *Der Spiegel,* (44). Available at: http://www.spiegel.de/wissenschaft/mensch/vererbung-und-erziehung-wie-eltern-ihre-kinder-praegen-a-325984.html.

Swinnen, S., Voordeckers, W. & Vandemaele, S., 2005. CAPITAL STRUCTURE IN SMEs: PECKING ORDER VERSUS STATIC TRADE-OFF, BOUNDED RATIONALITY AND THE BEHAVIOURAL PRINCIPLE. In European Financial Management Association.–Annual Meetings. pp. 1–40.

Szyperski, N. & Nathusius, K., 1999. *Probleme der Unternehmungsgründung,*

Tansley, A., 1935. The use and abuse of vegetational terms and concepts. *Ecology,* 16(3), pp.284–307.

Tax Foundation, 2013. Federal Capital Gains Tax Rates, 1988-2013. Available at: http://taxfoundation.org/article/federal-capital-gains-tax-rates-1988-2013.

Tax Foundation, 2011a. National and State Corporate Income Tax Rates, U.S. States and OECD Countries, 2011. *Tax Foundation.* Available at: http://taxfoundation.org/article/national-and-state-corporate-income-tax-rates-us-states-and-oecd-countries-2011.

Tax Foundation, 2011b. State Corporate Income Tax Rates.

taz, 2014. Sind Medien noch Vierte Gewalt? Available at: http://www.taz.de/!137953/.

Techcrunch, 2009. Amazon closes Zappos Deal. Available at: http://techcrunch.com/2009/11/02/amazon-closes-zappos-deal-ends-up-paying-1-2-billion/.

Techcrunch, 2011. Samwer brothers...Alando. *Techcrunch.* Available at: http://tech crunch.com/2011/02/18/the-samwer-brothers-make-a-killing-after-selling-their-fa cebook-stake-from-2008/.

Techcrunch, 2012. Yelp acquires Qype.. *Techcrunch.* Available at: http://tech crunch.com/2012/10/24/yelp-pays-50m-to-acquire-its-big-european-rival-qype/.

Telegraph, 2011. Ricardo. Available at: http://www.telegraph.co.uk/technology/ 8266718/Stefan-Glaenzer-Start-Up-100-judge.html.

The American, A Nation of Givers. *The American, The online magazine of the American Enterprise Institute.* Available at: http://www.american.com/archive/2008/march-april-magazine-contents/a-nation-of-givers.

The Economist, 2011. Another Digital Gold Rush. *The Economist.*

The Economist, 2013. Crash course. , (Sep 7th).

The Economist, 2006. The rise of the social entrepreneur. , (23.02.2006). Available at: http://www.economist.com/node/5517666.

TheFreeDictionary, 2012. Skills & Abilities. *TheFreeDictionary.* Available at: http://www.thefreedictionary.com [Accessed January 3, 2012].

The Tech, 2003. America and its Contradictions. *The Tech.* Available at: http://tech.mit.edu/V122/N64/col63basil.64c.html.

The Washington Times, 2013. HANSON: A nation of promiscuous prudes - America grooves on schizophrenic sexual morality. Available at: http://www.washington times.com/news/2013/apr/19/a-nation-of-promiscuous-prudes/.

The Whitehouse, 2013. Startup America. Available at: http://www.whitehouse.gov/ startup-america-fact-sheet.

The World Economic Forum, 2013. *Entrepreneurial Ecossytems around the globe and company growth dynamics,*

Thomson Reuters, 2014. *National Venture Capital Association Yearbook - Yearbook 2014,*

Timmons, J.A. & Spinelli, S., 1994. *New venture creation: entrepreneurship for the 21st century,* Irvin Press.

Timmons, J.A. & Spinelli, S., 2004. *New venture creation: entrepreneurship for the 21st century,* McGraw Hill, Irvin Press.

Timmons, J.A., Zacharakis, A. & Spinelli, S., 2004. *Business plans that work,*

Toren, M., 2011. Top 100 Entrepreneurs Who Made Millions Without A College Degree. *Business Insider*. Available at: http://www.businessinsider.com/top-100-entrepreneurs-who-made-millions-without-a-college-degree-2011-1?IR=T.

Turbotax, 2013. Capital Gains Tax. Available at: https://turbotax.intuit.com/tax-tools/tax-tips/Investments-and-Taxes/Guide-to-Short-term-vs-Long-term-Capital-Gains-Taxes--Brokerage-Accounts--etc--/INF22384.html.

Tyebjee, T.T. & Bruno, A.V., 1984. A model of venture capitalist investment activity. *Management Science*, (30(9)), pp.1051–1066.

Überla, J.D., 2011. VC in Deutschland.

UBS, UBS Geschäftspläne. Available at: http://www.ubs.com/ch/de/swissbank/business_banking/kmu/geschaftspl.html [Accessed March 27, 2012].

University of Portland, San Francisco Entrepreneurship Scene - show.aspx. Available at: http://www.up.edu/showimage/show.aspx?file=16782.

U.S. Chamber of Commerce Statistics and Research Center, 2005. Access to Capital: What Funding Sources Work for You.

US Congress, *Bankruptcy*, Available at: http://www.uscourts.gov/FederalCourts/Bankruptcy/BankruptcyBasics.aspx [Accessed September 10, 2012].

US Congress, 1958. *Small Business Investment Act of 1958_0*,

usistf.org, 2014. Appendix 1: The Experience of Six Small Successful Nations. Available at: http://www.usistf.org/wp-content/uploads/2014/03/The-Experience-of-Six-Small-Successful-Countries-Israel-2028-Appendix-1.pdf.

U.S. Small Business Administration, Business Plan Templates. Available at: http://web.sba.gov/busplantemplate/BizPlanStart.cfm [Accessed March 21, 2012].

U.S. Small Business Administration, 2012. Frequently Asked Questions. Available at: http://www.sba.gov/sites/default/files/FAQ_Sept_2012.pdf.

U.S. Small Business Administration, 2014a. Grants. Available at: http://www.sba.gov/content/grants-0.

U.S. Small Business Administration, 2014b. Loan Amounts, Fees & Interest Rates. *www.sba.gov*. Available at: http://www.sba.gov/content/7a-loan-amounts-fees-interest-rates.

Vasilescu, L., Business Angels: Potential Financial Engines for Start-Ups. *Ekonomska Istrazivanja/Economic Research*, (22(3)), pp.86–97.

Van de Ven, A.H., 1993. Predicting new venture survival:: An analysis of. *Journal of Business Venturing*, (8(3)), pp.211–230.

Venture Capital Magazine, 2013. Überblick über die deutsche Venture Capital-Landschaft. *Venture Capital Magazin*. Available at: http://www.vc-magazin.de/finanzierung/venture-capital/item/2555-%C3%BCberblick-%C3%BCber-die-deutsche-venture-capital-landschaft.

Verworn, B. & Herstatt, C., 2002. The innovation process: An introduction to process models.

Wagner, T., 2012. Creating Innovators: Why America's Education System Is Obsolete. *Forbes Magazin*. Available at: http://www.forbes.com/sites/ericaswallow/2012/04/25/creating-innovators/.

Wallstreet Journal, 2014. Andreessen Horowitz Raises Yet Another $1.5 Billion Fund. Available at: http://blogs.wsj.com/venturecapital/2014/03/27/andreessen-horowitz-raises-yet-another-1-5-billion-fund/.

Wallstreet Journal, 2007. Do Start-Ups Really Need Formal Business Plans? Available at: http://online.wsj.com/news/articles/SB116830373855570835.

Wells, W.A., 1974. *Venture Capital decision making*,

Whalley, K., 2014. Epigenetics: Early trauma alters sperm RNA. *Nature Neuroscience*, 15(6), pp.349–349.

White, L.J., 1966. *Medieval Technology and Social Change*, Oxford University Press.

Wikipedia, 2014a. Deutschland. Available at: http://de.wikipedia.org/wiki/Deutschland.

Wikipedia, 2014b. Kalifornien. Available at: http://de.wikipedia.org/wiki/Kalifornien.

Wikipedia, 2011. Liste der größten Unternehmen in Deutschland. Available at: http://de.wikipedia.org/wiki/Liste_der_gr%C3%B6%C3%9Ften_Unternehmen_in_D eutschland.

Wikipedia, 2014c. List of mergers and acquisitions by Google. *Wikipedia*. Available at: http://en.wikipedia.org/wiki/List_of_mergers_and_acquisitions_by_Google.

Wilkenson, A. & Kupers, R., 2013. Living in the Futures. *Harvard Business Review*, (Ma 2013).

Winborg, J. & Landström, H., 2001. Financial bootstrapping in small businesses:: Examining small business managers' resource acquisition behaviors. *Journal of Business Venturing*, (16(3)), pp.235–254.

Wirtschaftspedia, 2012. Käufermarkt und Verkäufermarkt. Available at: http://wirtschaftpedia.wikia.com/wiki/K%C3%A4ufermarkt_und_Verk%C3%A4ufer markt.

Wirtschaftswoche, 2013. Deutschlands Hidden Champions 2013. Available at: http://www.wiwo.de/unternehmen/mittelstand/markenranking-deutschlands-hidden-champions-2013/8955312.html.

Wirtschaftswoche, 2014. Neumacher. *Wirtschaftswoche.* Available at: http://award. wiwo.de/gwb2014/.

Woolley, J.L., 2011. Building the Infrastructure for Entrepreneurship in Emerging Domains of Activity. Available at: Building the Infrastructure for Entrepreneurship in Emerging Domains of Activity.

WordNetSearch, 2010. Success at WordNet Search. Available at: http://wordnet web.princeton.edu/perl/webwn?s=success&sub=Search+WordNet&o2=&o0=1&o7= &o5=&o1=1&o6=&o4=&o3=&h= [Accessed October 10, 2010].

World Bank, 2011. Gross domestic product 2010. Available at: http://siteresources. worldbank.org/DATASTATISTICS/Resources/GDP.pdf [Accessed March 16, 2012].

Worldbank, 2013. Market capitalization of listed companies. *data.worldbank.org.* Available at: http://data.worldbank.org/indicator/CM.MKT.LCAP.CD.

World Economic Forum & Booz & Company, 2011. *Accelerating Entrepreneurship in the Arab World,* Available at: http://www3.weforum.org/docs/WEF_YGL_Accelerating EntrepreneurshipArabWorld_Report_2011.pdf.

Wright, R.M., 1998. Venture Capital and Private Equity: A Review and Synthesis. *Journal of Business Finance & Accounting,* (25(5-6)), pp.521–570.

Würth, R., 2003. Entrepreneurship I+II.

Würth, R., 2001. *Entrepreneurship in Deutschland-Wege in die Verantwortung,*

Yahoo Finance, 2014. Market Cap of Yelp, Groupon,Facebook, ebay, amazon,. *Yahoo Finance.* Available at: http://finance.yahoo.com/q?s=YELP.

Zacharakis, A. & Meyer, G.D., 1998. A lack of insight: do venture capitalists really understand their own decision process? *Journal of Business Venturing,* (13(1)), pp.57–76.

Zeit-Online, 2013. Warum Frauen erfolgreicher gründen. Available at: http://www.zeit. de/karriere/beruf/2013-09/frauen-unternehmensgruendung-erfolg.

Zencke, P.D., 2010. SAP R3.

Zenthöfer, W. & Leben, G., 2008. *Körperschaftsteuer und Gewerbesteuer* 14., aktualisierte Auflage., Schäffer-Poeschel.

Zhao, H. & Siebert, S.E., 2006. The big five personality dimensions and entrepreneurial status: A meta-analytical review. *Journal of Applied Pschology*, pp.259–271.

Zutshi, R.K., 1999. Singapore venture capitalists (VCs) investment evaluation criteria: A re-examination. *Small Business Economics*, (13(1)), pp.9–26.